7/8/02

Sidney —

Here is a memento of Springhill — it tells the story of a special place to the S.R. Williamson family —

Tony Bonner was a playmate, class mate and close personal friend —

Love,
Pops

Springhill, Louisiana

THE CITY THAT PINE TREES BUILT

A Centennial History
1902–2002

CHARLES GARY BONNER

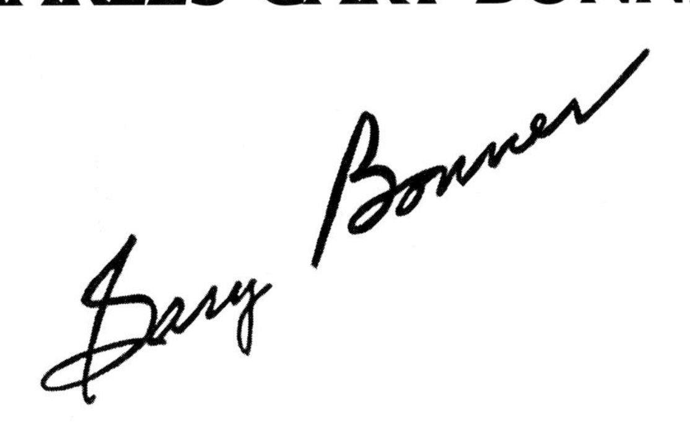

This publication is funded under a grant from the Louisiana Endowment for the Humanities, the state affiliate of the National Endowment for the Humanities. The opinions expressed in this publication do not necessarily represent the views of either the Louisiana Endowment for the Humanities or the National Endowment for the Humanities.

Sponsored by the City of Springhill, Louisiana

For CIP
information,
please access:
www.loc.gov

FIRST EDITION

Published in the U.S.A.
By Sunbelt Eakin Press
A Division of Sunbelt Media, Inc.
P.O. Drawer 90159
Austin, Texas 78709-0159
email: eakinpub@sig.net
website: www.eakinpress.com

1 2 3 4 5 6 7 8 9
1-57168-665-7

This book is dedicated to my **Family**:

. . . My parents, Flossie and Curtis Bonner, who taught family values by word and attitude and who were life-time residents of Springhill;

. . . My wife's mother, Lelia Harper, who accepted me into the family graciously and immediately and who was a long-time resident of Springhill. C. C. Harper died in 1953.

. . . My children, Andrea and Christopher, who have brought joy and hope to my life and who love to visit the city; and,

. . . My wife, Ann, loving lifetime partner and caring best friend, fun-loving native of Springhill in youth and responsible citizen of Springhill in retirement.

Memorial Dedication

The Rev. Dr. Gary Bonner's commitment to writing the history of Springhill sustained the final months of his life before his unexpected death. A distinguished Baptist clergyman, his and Ann's desire to return to their childhood home gave him a chance to reflect upon his boyhood experiences, on the stories of his family's early association with Springhill in the pre-IP days, and on the lives of longtime Springhill residents. Some of us were friends of Gary's—playing sandlot football and basketball, sharing high school times—and then in later life coming together to swap experiences and to rekindle our affection for Springhill. We applauded his decision to write this book and we are so very grateful that he managed to finish it.

Gary's last years, hampered by ill health, saw a heroic son of the Springhill community determined to record the history of the town he so loved and cherished. Ann, his children, his friends, and his readers all owe him special thanks for this labor of love and of lives and of life about and for Springhill. Dedicating the book to him is a fitting tribute to one who helped to shape the quality and aspirations of his generation of Springhill citizens.

—SAM WILLIAMSON, PH.D.

Charles Gary Bonner
1937–2001

Trees

I think that I shall never see
A poem lovely as a tree.

A tree whose hungry mouth is pressed
Against the earth's sweet-flowing breast;

A tree that looks at God all day,
And lifts her leafy arms to pray;

A tree that may in summer wear
A nest of robins in her hair;

Upon whose bosom snow has lain;
Who intimately lives with the rain.

Poems are made by fools like me,
But only God can make a tree.[1]

—Joyce Kilmer
1886–1918

Contents

Acknowledgments ix
Foreword xiii

Part One: The Forest Environment
Chapter 1: The Land—Physical Geography 1
Chapter 2: Numerous Species—Human Geography 6

Part Two: Early Growth—The Forest Grows 1810–1937
Chapter 3: Seedlings—Early Beginnings 13
Chapter 4: Saplings—Youthful City 21

Part Three: Rapid Growth—The Forest Expands 1937–1979
Chapter 5: The Greatest Pine—International Paper Company 37
Chapter 6: Great Pines—Economic Growth 50
Chapter 7: Great Pines—Educational Growth 65
Chapter 8: Great Pines—Government Growth 78
Chapter 9: Great Pines—Medical Growth 94
Chapter 10: Great Pines—Organizational Growth 110
Chapter 11: Great Pines—Recreational Growth 120
Chapter 12: Great Pines—Religious Growth 135

Part Four: New Growth—The Forest Renews Itself 1979–2002
Chapter 13: The Pines Fall—Economic Loss 157
Chapter 14: New Seedlings—The New Springhill 161

Part Five: The Forest Matures

Chapter 15: Stress in the Forest—Difficult Times 177
Chapter 16: Diversity in the Forest—African-American Culture 186
Chapter 17: The Neighboring Forest—Cullen, Sarepta, and Shongaloo 194

Part Six: Pine Cones on the Tree—Folklore and Biography

Chapter 18: Folklore in Springhill 211
Chapter 19: Biographies of Citizens 219

Part Seven: Reflecting on the Forest—Past and Future

Chapter 20: Historical Conclusions 243
Chapter 21: Unfinished Business 245

Appendixes

A—People of Northwest Louisiana. 247
B—Major Eras of Springhill History. 247
C—Buchanan's Empire 248
D—Historical Events in Springhill 249
E—Mayors of Springhill 250
F—Principals of Springhill High School. 250
G—Founding of Springhill Churches 250
H—Organizations 250
I—Cultural and Recreational Events. 251
J—Springhill Veterans of World War II 251

Notes 253
Sources Consulted. 259
Index 263

Acknowledgments

Following the Great Depression and World War II a citizen wrote,

> *As 1948 gains momentum in these post-war days, we find Springhill still building, still growing, and still groaning with sundry growing pains, which we are glad to say, are being eased gradually. Though the transition period has been at times painful, we feel that ultimately we will have built a bright and beautiful city with a bright and prosperous future. It is not what we may receive from our community, but what we contribute to it that counts.*[2]

With these simple words Kate Butler summarized this city built in the beautiful pine forests of Northwest Louisiana. She believed a review of the past would serve as an inspiration to face the future. Even in hard times, as well in as the prosperous ones, the citizens believed the city had a vital future. Former Mayor James Allen (1962–1974) summed up the prospects for the future when he said, "By reflecting on the history of Springhill, reviewing the facts, and observing what has been done, it is evident that this is a progressive, forward-thinking community. The close co-operation of all citizens has created this atmosphere in our city."[3]

This is the story of Springhill's rich history, written about individuals and events covering a period of one-hundred years, and about a beautiful and progressive community "built on the pine tree." Writing this history of my hometown has been a deeply rewarding experience, but a difficult task. The difficulty has been the decision-making process regarding what to retain in the book and what to leave out. Space limitations caused me to set aside some interesting data that I preferred to keep in the history. Nevertheless, I have attempted to tell the primary story by using biographies, human-interest experiences, and historical events. Furthermore, I have tried to capture the drama and spirit of the people who built this "shining city on a hill" through hard work, courageous leadership, and co-operative community spirit. You, the reader, will be the final judge of the extent this goal has been met.

Primary sources such as interviews and original documents were used as extensively as possible. Secondary sources such as second-hand stories of others and book interpretations of historical events were used when appropriate.

Numerous people have contributed to the writing of this book. I wish to thank all of them from a grateful heart. However, some have played such an important role that their names should be mentioned. I thank Mayor Johnny Herrington for his recommendation to have a Centennial Celebration, for his generous support and for his wise counsel. Likewise, I thank Jan Willis for superb leadership in the total project—co-ordinating the work of numerous people as the program developed and offering encouragement as every detail was considered. The Centennial Planning Committee has my heartfelt appreciation for the leadership role all of them assumed in all areas of the celebration. Members are Dr. A. C. Higginbotham, Chairman, Francis Eason, Eluida Flanakin, Jimmie Sue Murph, Joe Curtis; Rev. Randall Murphy, Ray Huddleston, Fannie Moore, Caldwell Colvin, and Diane Stephens.

The History Book Committee worked many hours collecting pictures for the book, assisting in research, working on the Index, and supporting and encouraging the author. I thank them all: Lois Fritz, Chairperson, Dr. Kathryn M. Benson, Pat Knesel, Cindy Hall, and Ann Bonner.

Carole Castleberry deserves our special recognition and appreciation for writing the grant application to the Louisiana Endowment for the Humanities.

The names of the individuals who contributed research information are legion. I offer my sincere gratitude to them for their contribution to the book. They are Howard Beaty, Georgia Branton, Sim McDonald, Tom Craig, Jean Erwin; Richard Noles, Evelyn Hardy, Elizabeth Brewer, Dewey Williams, Page Williamson; Charles McConnell, Rev. Kenneth Everett, Rev. Randall Murphy, Sherrel Smith, Jonathan Washington, Betty Rhynes, and Dr. Sam Holladay.

Dr. Kathryn M. Benson read the earliest manuscript and made constructive recommendations about the content and style. Her ideas were seriously considered and acted upon, but the final wording and the historical interpretation within the text are the author's responsibility. I offer my thanks to Dr. Benson for her excellent work.

Dr. B. H. Gilley assisted the Centennial Committee by serving as Project Consultant. This included consultation with the author about the Springhill history book. His scholarly observations and suggestions were greatly appreciated.

A special thank you is extended to my former secretary in Huntsville, Texas, for typing the final manuscript. Peggy Mathis has always been professional in her work. She was no less in this endeavor. Her work is superb.

Most of all I am appreciative of and thankful to my wife, Ann, who offered suggestions and critiqued the manuscript as it developed. She was especially helpful about the proper use of English grammar. I am grateful for her support of me and her patience with me during the year and one half of writing the book.

From the beginning the author has had the freedom to write without censorship. Nevertheless, I sought historical data, human interest stories, and biographical information from numerous sources. Many of these suggestions are included in this book. A hearty thank you goes to these deserving contributors.

This volume contains a number of vignettes that provide insight into the lives of early leaders in Springhill. These brief biographical sketches make the book more than a presentation of facts. It becomes a historical pilgrimage of personalities who built the city on the economics of the pine tree. Thank you, families, for providing information about loved ones who have made a difference in the history of Springhill.

The pictures of historical events and early citizens are priceless. A very special thank you is given to all persons who brought photographs that are printed in the book. These photographs will be a valuable resource for future generations who wish to study Springhill's history. A special thank you is given to Lois Turner Fritz, Evelyn Krouse Hardy, Andrea Bonner, Bettye Benton Rhynes, R. O. Machen, Jr.; Robert C. Bryan, Barbara Barnard Bryan, James J. Holland, David Jeane, A. C. Higginbotham; Gary Bonner, Ann H. Bonner, Samuel R. Williamson, Eluida H. Flanakin, D. C. Wimberly; Brian E. Driskill, Wm. Charles Park, the negatives collection of C. J. Provost, Chase Studio of Washington, D.C., Revelle Studio of Huntsville, Texas, the University of Georgia Press; and Grisham Studio of Magnolia, Arkansas, and Springhill.

Mack Memorial Libary loaned its album of Fritz Photographers prints. Derek Melancon at Springhill Medical Center scanned to CD its library of Springhill photos originally provided by Martha Machen Horne and Herbert Owen Park. The Springhill High School mural, Alma Mater, and championship sport photos are from yearbooks.

Springhill *dot* Net provided supplies and time for Joshua Hanson to scan all selected photos to CD for this Centennial book project.

Foreword

The grinding of sawmill machinery on a summer night, the pungent odor from the paper mill on a wet evening, the rumble of heavy oil rigs moving through town all have characterized Springhill, Louisiana during the twentieth century. The interaction of people, industry, natural resources, and recurrent bouts of hope and despair have defined the life and fabric of Springhill. But it could have defined the life of Barefoot, Louisiana save for the arbitrary action of J. F. Giles at Christmas, 1896 when he simply put up a sign and gave the sawmill community a new name: Springhill.

From that point forward and after incorporation in 1902, Springhill has seen modest growth, economic depression, boom town prosperity, sustained civic success and then economic hardship, and now a resurgence of local pride and hope as the economy diversifies and ambitions reassert themselves.

Gary Bonner has chronicled, in a labor of love and pride, the life of our hometown through the twentieth century and into the next. Deftly weaving the theme of pine trees with that of oil, Gary has examined the twin economic impacts of these two natural resources in his history of this "New South" town. With careful attention to Springhill's geological and geographic inheritance, he has examined the separate strands of community life: the economics of the pine tree seen in the saw mill and paper mill, religious life, medical care, athletic endeavors, race relations, leisure activities, and community organizations. Vignettes of individuals give the reader a sense of joy and association with those of the past and the present. Carefully noting the role of key pioneer leaders and pioneer families, Bonner has shown how successive generations have labored to enhance Springhill's community life and economic potential. His history assesses the role of organizations, individuals, and external economic forces—the arrival of the Pine Woods Lumber Company to the departure of paper productions by International Paper.

His sensible conclusions and suggested agenda for the future of Springihll reflect a realistic evaluation of the things that have gone right and wrong in the life of the town in the past, while prudently offering some thoughts for the future. The future leaders of Springhill will want to keep a copy of this volume nearby. And all readers, even those familiar with the history of Springhill, will learn much from this wonderful contribution to the history of a major center of life in North Louisiana.

—Sam Williamson, Ph. D.

PART ONE

The Forest Environment

Chapter I

The Land–Physical Geography

Geological Pluralism

Geological time. The author experienced two happy events that are etched in his memory from his days as a young lad. Both experiences represent two natural resources basic to the Louisiana economy and to the history of Springhill.

During the Christmas season my father, Curtis, and my mother, Flossie, would take my brother, Jerry, and me into the woods east of the present-day civic center and city hall. There we would discover and cut a perfect little pine sapling to serve as our Christmas tree. The family would take it home, decorate it with strings of popcorn, aluminum icicles, and inexpensive glass ornaments, and then enjoy it immensely until the Christmas season concluded about December 31st.

The second experience occurred during my teen years in the summer time. My friend, Jimmy Chadwick, asked his father, Ray, to employ us in the oil fields near Springhill during our summer vacation months. The manual labor was hot and dirty, but it gave us an opportunity to make money for the forthcoming school year. We delighted in it although the huge oak timbers were covered with mud and were most difficult to move around to build roads from the wellhead to the outside gate. By the end of the summer we had become accustomed to the work, but then it was time to return to school.

Both experiences mirrored the symbols of Springhill history, the pine tree and the oil field. They represent the plentiful natural resources that were used as the basis for the establishment and growth of this beautiful small town nestled among the pine trees of Northwest Louisiana.

Springhill dating. Springhill's written history begins in 1896, but resources for growth and development have been forming for approximately 70,000,000 years according to geologists.

Historical forces. Geological forces have produced rocks, domes, and folded layers of earth over these millions of years. They contain oil and soil used by early settlers and present-day residents to build an economy for their families. Meteorological forces have provided long growing seasons, a semi-tropical climate good for plant growth, and rainfall adequate for the growth and nurture of plants. Springhill is located in an area for perfect growth of the pine tree.

Geological time. Across millions of years of geological history various types of rocks have been formed through the compression of sediments, the intrusion or extrusion of magma (molten rocks from the earth's center), or the heating and pressurization of existing rock.[1] Geologists study these structures of the earth through a system called "geological time" which is divided into five principal "eras": Archeozoic, Proterozoic, Paleosoic, Mesozoic, and Cenozoic. Practically all of Louisiana dates from the most recent era—the Cenozoic—and is, therefore, young as far as geological time is concerned.[2]

Major landforms. Several types of landforms are located in the state of Louisiana. The hills and uplands are in the western and northern part of the state which include Springhill.[3] These hills are associated with domes, upward folds and layers of rock which have created pools of oil in the seams. Other landforms in Louisiana include flood plains, bluff lands, prairies, and coastal marshes. Terraces are found also in the state of Louisiana. These were formed over geological time frames when shifting global water levels caused rivers to first deposit, then to erode away, alluvial material. The pine flats in Northwest Louisiana are part of these ancient terraces.[4] Springhill is located in the midst of these terraces and in the hills that have both soil and climate conditions that are conducive to pine tree growth and oil development.

Upland Frontier

Trackless wilderness. The hill country between the Ouachita and Red Rivers was a trackless wilderness in 1800. The entire area was an unbroken forest through the first two decades of the nineteenth century.[5] It was impossible to get a wagon through the forests. Horseback riding along the Indian trails would sometimes be successful. Usually a man on foot could walk into the forests along the trails, but he often encountered a mass of tangled underbrush that prevented him from continuing the journey.

Dr. Philip Cook, history professor at Louisiana Tech, quoted an early historian who wrote about the tangled underbrush that prevented settlers from entering the area to build their homes and plant their crops. This early history document of Northwest Louisiana described the country side as it appeared to the first settlers:

> North Louisiana, at this time, was covered with a dense mass of brush-wood and interlacing vines—home of the wolf, the bear, and the panther. Numbers of horses and cattle, the progenitors of which had wandered from the inhabited sections of the territory to this wilderness, ran free and wild. The few early settlers that ventured into these wild regions had to fairly hew their way, for only a few devious trails and paths were to be found. But gradually, settlement clearing increased, and from these clearings and the camps of the hunters, fires broke out sweeping over all the land, killing the tangled undergrowth or brush-wood, even destroying the

foliage of lofty trees. In the following years fires again raged, consuming all the dead and fallen rubbish that then encumbered the ground.[6]

The fires. These fires, which dated back to 1824, had a profound effect on the appearance of North Louisiana. Dr. Cook wrote,

> Being thus relieved of its heavy undergrowth, in its place forest grass and switch cane sprang up, and in one season a mantle of green covered the nakedness of the earth. Then all North Louisiana appeared as an immense park, diversified with vast openings, and vistas most enchanting.[7]

Hunter's Paradise. Wild game of every kind increased rapidly, especially deer and turkey. Buffalo came up from the Attakapas prairies of Southwest Louisiana. In a few years North Louisiana became known as the "Hunter's Paradise." Today the concept has been expanded. Northwest Louisiana is known as the Sportsman's Paradise on maps, in brochures, and in private conversations. The Louisiana Tourist Commission identifies the entire Northern Louisiana area as a Sportsman's Paradise. The name was given to the area in the early decades of the 1900s.

Transportation. Transportation was an important factor in the growth and development of Northwest Louisiana. The first settlers depended entirely on Indian trails and waterways. Land travel could be accomplished only on foot or in limited areas by horseback.[8]

Not only was there an absence of roads, but the Red River was encumbered with the Great Log Raft. The Red River log jam was a barrier which kept trading posts from being established north of Natchitoches for a long period of time. Without roads and navigable waterways it was difficult for people to enter the region. There was no way to market goods that might be produced by the early settlers.

In 1828 the U.S. Army built Military Road No. 11 northward from Natchitoches to Fort Jessup in Arkansas Territory. The other main route was built east and west from present-day Monroe to Shreveport.[9] Congress appropriated $25,000 in 1828 for Red River navigation improvement. The state legislature added another $15,000 for the improvement of Loggy Bayou.[10] These two actions opened Lake Bistineau as a major artery for transportation in North Louisiana and facilitated a mass settlement of the Northwest Louisiana uplands during the following two decades.[11]

John Murrell, the first English-speaking person born in Northwest Louisiana, lived in what is known today as Claiborne Parish. He described this new attitude about frontier life in Northwest Louisiana:

They lived more like a band of brothers than the gathered-up fragments from different states and nations. A crime was rare. When a stranger came who did not identify his standing as a good character, he was waited upon and given to understand that the people governed, and that he had better be honest or move.[12]

Forests of Louisiana

Early forests. At the beginning of the eighteenth century, eighty-five per cent (twenty-two million acres) of Louisiana was covered with Southern pine. These great forests were mentioned in the writings by early travelers such as LePage du Pratz,

Charlevoix, Bossa, and Bertrum. Tall pines were used for ship's masts, construction timber, and tar for the Indians.[13]

After statehood. Lumbering was conducted on a small scale. It was confined to areas adjacent to lakes and streams that were used for transportation of their products. This was the situation in 1880 when a timber survey was taken as part of the Tenth Census. A map published at that time shows the vast pine lands of Louisiana had a very small amount of cut over land.

After Civil War. At the time of the Civil War, the pine forests of central and northwest Louisiana were described as "those unending pine woods," and "sad and monotonous pine forests" by Felix Pierre Poche. These were the virgin forests that would be viewed by the timbermen in the late nineteenth century as potential wealth-builders.

Nineteenth century. In the nineteenth century timbermen moved into the virgin forests to construct sawmills and railroads. Springhill was the recipient of this economic activity when, in 1896, James Buchanan bought land and constructed a sawmill in Barefoot, Louisiana. The stage was set, the props were in place and the actors were coming on stage. The great drama was beginning.

Oil helped fuel Springhill's economy.

The forest—early basis for Springhill's growth.

West side oil production.

Pine trees, full of new growth.

Chapter 2

Numerous Species–Human Geography

During junior high school days the author visited a friend who lived on the road to the paper mill ponds. George Smith's house was located south of the present high school approximately one-half mile from the intersection of 7th St. N.W. and Church Street. South of George's house was a small creek running east and west. On one of my visits we walked through a small field near the creek. The field, which had been recently plowed, had dozens of arrowheads on top of the ground. We placed a few of them in our pockets without realizing the importance of the discovery. This site was probably the location of a Caddo Indian village or perhaps a camp site used for an extended period of time by the Caddoans.

Tens of thousands of years ago, when the world was in the midst of the Ice Age, the first humans made their way into North America. At that time an extensive land bridge connected Siberia to Alaska which is now called the Bering Stait. People from Asia used this route for their passage into North America. Over hundreds of generations, nomadic people spread throughout southern North America, Central America, and South America. By 10,000 B.C. the first Indians lived in the Southeastern United States. The pre-historic era in Louisiana begins with these first inhabitants.[1]

Northwestern Louisiana was occupied by Caddo Indians during the period of early Spanish, French, and American contacts. By using a combination of history and archeology the Caddo story can be traced back for a thousand years. Indian artifacts discovered in the plowed field south of the present high school on 7th Street N.W. were pieces of evidence demonstrating that Indians lived in the Springhill area hundreds of years ago.

Pre-historic. Many Indians probably appeared first in what is now Louisiana about twelve thousand years ago.[2]

A chart in the Appendix A of this book provides an overview of the people in Northwest Louisiana from 10,500 B.C. to the present.

Caddo Indians. About 800 B.C. people living in Northwestern Louisiana had developed close ties with people in Southeastern Oklahoma, Northeastern Texas, and

Southern Arkansas. From this region emerged the Caddo culture.[3] They lived in small villages near streams, grew food in gardens, celebrated festivals and ceremonies, fashioned jewelry, and decorated pottery. The Caddo Indians lived along the Red River basin, the Ouachita River basin, Bodcau Bayou near Bellevue, and in a large area near the present city of Natchitoches.

Evidence of the Caddo Indian culture has been located at the Montgomery site at the Springhill airport in upper Webster Parish. The people apparently lived here long enough for their thatched roofs and clay-daubed houses to have been repaired a number of times. Residues of gathered or hunted food stuffs are present: hickory nuts, acorns, persimmons, mussels, turtle, fish and deer bones. Their pottery ranged from rough culinary or storage pots to nicely engraved bowls and red-surfaced or engraved bottles.[4] Archeologist David Jeane of Springhill was involved in the excavation of this site.

What do the Indians have to do with Springhill history? Why are they included in this book? Because they lived here! Caddo artifacts have been found within the city limits. But more importantly, the Caddo Indians are part of the human geography of the area. Native Americans make up one element of the human species that have historical ties to the town site.

Explorers

Spanish. Spain pursued an aggressive exploration policy of the Western Hemisphere. Alonza Alvarez de Pineda (1519), Panfilo de Narvaez (1527), and Hernando de Soto (1539) led expeditions from Florida to the Mississippi River traveling as far north as Missouri and as far west as Arkansas.

Spanish explorers came close to the pine forest. They encountered the Caddo Indians in Arkansas. A land speculator, Manuel O'Garte of Mexico, received a Spanish land grant in the late 18th century which included about 4,000 acres between what today's citizens know as the paper mill area and Bodcau Bayou, and on into Bossier Parish.

French. LaSalle descended the Mississippi in 1682, reaching its mouth in April 1684. He named the territory "Louisiane" (Louisiana) after King Louis XIV of France.[5] On April 9 he took formal possession of the Mississippi basin.

Other explorers under Henri de Tonti (1682), Pierre Le Moyne Sieur d' Iberville (1698), and Louis Jachereau de St. Denis (1698) searched the Mississippi and Red Rivers but established nothing in Louisiana. One expedition by Jean Baptiste Benard de La Harpe departed New Orleans in 1718 to establish a trading post on the Red River among the Caddoan Nassonite Indians. On February 20, 1719, he set out again on the Red River, reaching present-day Texarkana.[6]

French explorers also came close to the Springhill area, but they established no forts or villages. French influence is evidenced in the South Louisiana area in the form of language, food, music, and law. Citizens with names such as Manuel, Bergeron, Barberousse, Jandebeur, Soileau, Breaux, and Troquill dot the telephone directory reminding us that the French culture has made its mark in Springhill.

American. After the Louisiana Purchase was negotiated in 1803, President Thomas Jefferson sought early exploration of the new territory. While Lewis and

Clark journeyed overland in the North, Jefferson made new plans for the South, particularly the Red River. Dr. John Sibley, a physician, explored the Red River in 1803, a year after settling in Natchitoches.

The American President and the American explorers were interested primarily in the Red River basin. These explorers came close to the Springhill region on their journeys. However, the greatest change came in Northwest Louisiana when the early settlers migrated in two waves, one from Tennessee and a second wave from the southern tier of southeastern states. The first wave came from Middle Tennessee and the second wave came from the Carolinas, Georgia, and Alabama across the Mississippi Territory into Southern Arkansas and Northern Louisiana.

Anglo-Saxons

Germans and English. The settlers in the hill country of North Louisiana, including what became Webster Parish, came from various places. Some Germans settled near Minden in 1720. They were from St. Charles and St. John the Baptist Parishes. The English migrated from the colonies across the Appalachian Mountains to Tennessee. From there the settlers came down the Cumberland River to the Mississippi River to the Red River and up past the Great Log Raft into Claiborne Parish. Claiborne Parish, when it was created in 1822, contained the territory now comprising Claiborne, Bienville, Bossier, Webster, and portions of Red River, Jackson and Union Parishes.

The eastern part of the hill country was settled earlier than the western side before the Louisiana Purchase was negotiated. The first settler was John Honeycutt who obtained a Spanish land grant sometime during the 1790s. He moved to present-day eastern Union Parish. A few years later a second Anglo-Saxon family, the Feazels, moved into the area of the present town of Downsville. They had come from Tennessee with "a house full of girls," and according to legend, Honeycutt, upon seeing them, bluntly stated, "I'll take this un for my wife."[7]

When Louisiana became a state in 1812, a group of families from Chester County, South Carolina, came up the Ouachita River to Fort Miro near present-day Monroe, Louisiana, and proceeded westward into the eastern and central part of the hill country. These families were named Huey, Colvin, Sims, and Rainey.[8] Daniel Colvin moved into an area a few miles north of Ruston in the center of Lincoln Parish. Originally called Collinsville, it was renamed Vienna in 1837. The settlement flourished and became the principal town of North Central Louisiana. Today the Colvin family constitutes the largest family clan in the uplands.[9]

Others who came to the northern hill country were the Abraham Pipes family who settled near Choudrant and the Jesse Hagler family who settled north of Chatham in 1815.

Migration of 1817. In the winter of 1817 during the boom times known in American history as the Era of Good Feelings, Colonel William Clark, who had made a previous expedition to the Arkansas Territory and the upper Red River Valley, gave people of Middle Tennessee a glowing description of the "Long Prairie" country and the Great Red River Raft. This set off a wave of migration from Middle Tennessee that would have great consequences for Northwest Louisiana. Numerous families left

Tennessee via keelboats bound for the lands along the Red River in Arkansas. The area turned out to be malarial, so many newcomers fled to a more healthy environment. The Northwest Louisiana uplands was nearby and unsettled, providing an attractive alternative to former Tennessee natives.

When the sawmill was constructed in Barefoot, Louisiana, in 1896, a few poor farmers were found in the area. Perhaps they were settlers who came from Tennessee around 1817 during the Era of Good Feelings, or perhaps they migrated during the Panic of 1819 which caused an economic depression that lasted six years. Farmers throughout the nation were ruined, so many of them moved west to get a new start. Some of these people could have been the population base for the future village of Springhill.

The theory that the North Louisiana hill country region was probably settled by families with previous connections in Middle Tennessee and the Southeastern part of the United States is reinforced by data from the U.S. Census reports of 1850 for parishes of the region. It indicates that a large proportion of the older residents were born in Tennessee or in the Carolinas. The typical migration pattern was for Carolinians to first move over the mountains to Middle Tennessee and then several years later float down the rivers to North Louisiana, often after a short stay in Southern Arkansas.[10] The other wave of migration came from the Carolinas into Georgia, into Alabama, across the Mississippi Territory into Southern Arkansas and Northern Louisiana. The author's ancestors made this journey.

Slaves

It became evident that the rich soils of Louisiana could produce profitable crops, especially indigo and tobacco. It also became evident that few Europeans or Indians were willing to undertake the hard work necessary to clear the fields and plant the crops. The solution was found in African slaves. Five hundred arrived in 1716, and many more came in 1718 when the Company of the West assumed control of the Louisiana colony under Bienville's direction.[11]

Caddo Indian engraved bottle ca. 1600 A.D., found near Springhill Airport.

Indian points from Site T in March 1980.

Points from Site B, wastewater basin of paper mill.

Points from Site C, March 1980.

PART TWO

Early Growth–

The Forest Grows

1810–1937

Chapter 3

Seedlings–Early Beginnings

"Things born right grow best." This statement expresses a viewpoint of a popular philosophy among citizens. Taken from the agriculture world, it means a good beginning usually produces success. Seeds planted in good soil and introduced to a proper climate grow well.

The author observed this truth when he, his wife Ann, his two children, Christopher and Andrea, and his father, Curtis, planted 1,300 pine seedlings on twenty acres of land near the town of Springhill in 1974. It was a cold, wet December week near the holiday season when the pine seedlings were placed in the ground, one seedling at a time. The job required two days. The children were young and not too willing to work hard in the misty atmosphere and cold temperature. Nevertheless, we completed the task. In a few years the seedlings became saplings, and the saplings became young pine trees. Because the author and family lived in Texas, we did not see the trees often. However, when we did return to Springhill for holiday seasons, we would make a journey into the countryside to observe the trees. How they grew! Rapidly, tall, and beautiful. They would become a cash crop at some later year. These pine trees fulfilled the principle that "things born right, grow best." Cool temperatures, plenty of moisture, and winter sunshine got them off to a good start. Likewise, the early history of Springhill had a good start!

Migration Infuences

Civil War 1860–1864. The Civil War reduced the South to rubble. In the battles 258,000 white Southerners lost their lives and 200,000 were wounded. Cities were gutted, farms were burned, livestock was butchered, bridges were dynamited, ferries were destroyed, and railroads were dismantled. Industries and small businesses were forced to close their operations. Poverty, disease, and starvation spread into large sections of the country. Bands of thieves roamed the countryside creating fear and hostility among the citizens of the South. Added to this devastation was the political program of Reconstruction that was imposed upon the Southerners. Corruption was rampant. The suffering of the people continued as a result of this program, although it was intended to rebuild the South after the war.

It was 1869 before a record crop flushed the South with a wave of Northern capital. Small farms popped up, retail stores appeared in towns, and railroads expanded to meet the growing demands for travel. The good economic times following the Civil War pulled at the Southerners to join in the new opportunities. Migration into the Southern forest lands increased.

Golden Age of Lumbering. To the Northern business man's eyes the South was a rich resource of timber and cheap labor. Also, the lumber industry had moved from Pennsylvania and New York to the Great Lakes region in the 1860s and 1870s because of their cut and move policy regarding the forest. Now the Great Lakes region was depleted of timber for their sawmills. The companies began their move to the South. Migration of families accompanied the workers when they moved with the sawmill companies.

The repeal of the Homestead Act of 1876 by the United States Congress opened up millions of federal acres for private transactions. Over 47 million acres of land in Alabama, Florida, Louisiana and Mississippi were for sale. Railroad companies joined the cheap land opportunity. This, too, encouraged migration to the Southern states.

Early Migration Patterns

From Middle Tennessee. Colonel William Clark (Chapter 2) in 1817 made a trip into the Arkansas Territory. He carried glowing reports to the people in Middle Tennessee during the Era of Good Feelings. This report set off a wave of migration via keelboats into the Arkansas Territory and eventually into Northwest Louisiana. The Panic of 1819 occurred at this time. It created a migration of poor farmers seeking a better life. Northwest Louisiana had a population increase from former Middle Tennessee people.

From Southeastern United States. The second migration pattern began in the Carolinas, Georgia, Alabama and Mississippi. People traveled in small groups across the Mississippi Territory into Arkansas and Louisiana.

Early Settlers[1]

Isaac Alden 1811. The first known English-speaking settler in the Northwest Louisiana hill country was Isaac Alden who came from the North in 1811. He traveled through New Orleans into Northwest Bienville Parish (now Webster Parish). He settled eight miles northeast of present-day Minden and owned a water-powered sawmill, a grist mill, and a blacksmith shop.

John Turner Sikes 1814. The Sikes family settled on the east bank of Dorcheat Bayou at a crossing used by the Indians. John Turner Sykes had migrated from

Liverpool, England. Sikes and sons built and operated a farm and ferry known today as Sikes' Ferry. The Sikes family and the Alden family were the only known English-speaking inhabitants in the hills of Northwest Louisiana when several other families arrived in 1818.

Abraham Pipes. The Abraham Pipes family settled near the town of Choudrant in the Central Hill Country in approximately 1815.

Hagler family. This family placed their roots down near the present site of Chatham.

Tennessee farmers 1817. These were the farmers who came to Northwest Louisiana because Col. William Clark gave glowing reports to the citizens of Middle Tennessee about beautiful, fertile land. (See earlier paragraph).

Samuel Monzingo, *William Farmer*, *J. A. Byrnes*, *Joseph Murrell* and *Thomas Neal.* These families were said to have lived in the Springhill area around 1850. Samuel Monzingo and Thomas B. Neal purchased land in the southwest quarter of the north-east quarter of Section 11, Township 23 North, Range 11 West early in the year 1860. This is the location of the present First Baptist Church.

James Wise's sons 1850s. Several sons of James and Permelia Wise settled near the Indian village (present day Shongaloo). It became known as Wiseville. The first post office was established in 1857. Dr. Giles James Wise practiced medicine in the Indian village and in Wiseville.

Sarepta Carter 1868. The name of the town of Sarepta came from the name of Mrs. Sarepta Carter. She presented a pulpit Bible to the church when they named the church after her, Sarepta Church. The young village was developing at this time, so it became known as Sarepta.

Hugh Jacob Coyle 1878. Hugh and his wife Louvenia Braley Coyle settled on Pecan Street in Clifford (Cullen) where he operated a merchandise store and farmed cotton. In 1937 International Paper Company purchased some of the Coyle estate to build the paper mill.

William Harrison 1880. Doyles Crossroad (Shongaloo) was settled by William Harrison in 1880. He nailed a sign that read, New Shongaloo, to the door of his store on the northeast corner of the crossroad. Now Old Shongaloo and New Shongaloo existed together.

J. B. Rowland 1886. His father helped build the L & A Railroad from Stamps, Arkansas, to Springhill, Louisiana. In an interview with the *Springhill Press and News Journal* in 1979 Mr. Rowland recalled his eyewitness account of the building of Bodcaw Lumber Company in Barefoot, Louisiana (Springhill). He was the third and only graduate from Springhill School in 1912.

A.J. McDonald 1900. Dr. Andrew Jackson McDonald was the first licensed doctor to practice in Springhill. He rode a horse from Shongaloo to Springhill in order to practice medicine in the town in 1898–1899. He moved permanently to Springhill in 1900.

Rupert Butler 1917. Kate and Rupert Butler moved from Redland on the Plain Dealing highway to Springhill in 1917. He practiced medicine on the second floor of the bank building located at the corner of Ensey and Giles streets and on the patients who worked for Pine Woods Lumber Company. The Butler Memorial Health Center in Springhill is named after this beloved physician.

Influential Forefathers[2]

William Buchanan. James Wortham Buchanan, a bookkeeper by occupation, took his family from Shelbyville in Bedford County, Tennessee, to Southern Georgia near his wife's ancestral home. The year was 1862. The family later returned to Tennessee to Decherd, a small community a few miles from Winchester, in 1866. William Buchanan was seventeen years old.

Biographers are not quite certain why Buchanan joined the migration to the South, except for the fact that his entire life had been shaped by social winds blowing around his family. He had an impulsiveness driven by great ambition. Early in his life he exhibited characteristics of a private nature. Buchanan was, in fact, an independent, competitive individual. These traits would lead him to practice maximum control of his business affairs in later years.

Whatever his motivation for leaving home, when he left Tennessee in 1869, Buchanan carried with him a small portable sawmill. His destination was Forrest City, Arkansas. In 1873 he traveled to Texarkana, Arkansas. There he helped Joseph Ferguson who was a wealthy landowner in Texarkana. His friendship with Ferguson was probably the deciding factor in stopping his westward migration. It was here that he went into the sawmill business.

Three miles from the Sulphur River is the place where Buchanan constructed his first sawmill. The Panic of 1873 created a depression for six years that adversely affected his business. Furthermore, heavy rains flooded his timber making it impossible to get his logs to the sawmill. These business setbacks caused him to dream of wealth in other places. In 1877 he traveled to Leadville, Colorado to seek his fortune in silver mining. In 1880, after a brief time, economic hardships and his family's encouragement lured him back to a sawmill in Buchanan, Texas. By 1888 he finally exhausted his timber reserves on the Sulphur River, so he moved again to Texarkana. There he made friends with some of the wealthy businessmen who included G. W. Bottoms, E. W. Frost, and C. T. Crowell. These friendships would have a lasting effect on Buchanan's life.

In 1887 Crowell initially purchased the Bodcaw Lumber Company in Stamps, Arkansas (the company was spelled with a "w" although the bayou is spelled Bodcau.). When he tired of the lumber industry, Crowell sold the mill to Buchanan. Little is known about the transaction. Documents finally appeared on January 14, 1889, that recorded some of the stock transactions.

Stock in the mill was juggled, but it ended up in the hands of J. A. Buchanan, W. T. Ferguson, and William Buchanan. No record exists of the stock and leadership maneuvers, but it was clear that Buchanan was in control of the Bodcaw Lumber Company at the end of 1889.

The mill would bring wealth to Buchanan, yet he wanted more. His driving ambition and his desire for more wealth led him to purchase timber and build another sawmill in Barefoot, Louisiana, in 1896. This business initiative became the beginning point of the history of Springhill. William Buchanan would eventually own seven sawmills from Stamps, Arkansas to Jena, Louisiana.

The L&A Railroad was another one of Buchanan's business enterprises. A major concern of Buchanan was getting logs to the mill and manufactured lumber to the markets. Railroads were the answer, but the rail lines Buchanan observed were dead ends running only short distances. He was aware of the national railroad growth and the impact it could have on the transportation of lumber.

He looked at a map and saw a left leaning V formed by the junction of the Red River in Northwest Louisiana, the Ouachita River in Northeast Louisiana, and Alexandria in South Central Louisiana. Stamps, Arkansas, was located in the top left hand corner of the V on the map. Buchanan realized that if he could run a railroad from the top to the bottom of the V he could control all the area within it. It would be a simple thing to send branches out to Shreveport and New Orleans giving him access to cities and ports of the world.

Buchanan began pushing the rail line southward from Stamps, purchasing timber as he moved the line deeper into Louisiana. As he built the railroad, he built sawmills along the way. Springhill was one of them. Others were Minden, Stamps, Selma, Trout, Tall Timbers, and Jena. His empire now extended from Stamps, Arkansas, to Vidalia, Louisiana, on the Mississippi River.

The L&A Railroad was incorporated in 1901. When he opened his seventh sawmill in 1912, William Buchanan was sixty-three years old. His railroad was a commercial success carrying both passengers and freight, covering many miles, and employing 5000 people. His seven mills were shipping an average of 300 million board feet annually. Springhill was a direct result of Buchanan's sawmill and railroad expansion from 1896 to 1912.

J. F. Giles.[3] The other business man who greatly influenced the founding and development of Springhill, Louisiana, was J. F. Giles. He came to Barefoot, Louisiana, to be manager of the commissary and assistant to Mr. Harris, manager of the sawmill.

One of the first things Giles did as manager of the commissary was change the name from Barefoot to Springhill. Three accounts of the name change from Barefoot to Springhill have been preserved. All of them are essentially the same.

The first account is from Archie Mayor's biography of William Buchanan. He wrote of the event in this manner.

> A few days before Christmas, J. F. Giles, a Buchanan executive, pulled the sign 'Barefoot' off the store's front door, wrote 'Springhill' on the reverse side, and nailed it back up. The company had come to stay.[4]

A second account of the name change is recorded by Tommie O'Bier. It reads as follows:

> Mrs. Mullins, 'Uncle Buck Smith', Mrs. Twitty, Mrs. Hoyle, and Ralph Ensey got with Jimmy Giles and said, "We've got to rename this town. Barefoot is no fit name.' After some discussion they came up with the name 'Springhill' because of the hills and many springs. A few days before Christmas, 1896, J. F. Giles pulled the sign marked 'Barefoot' off the store's front door, wrote 'Springhill' on the reverse side, and nailed it back up.[5]

Mrs. A. B. Rowland, who lived in the village at the time Bodcaw Sawmill and commissary were constructed, wrote a letter that gave a vivid account ot the name change. This is the third description.

> I landed in Barefoot Station in November of 1896. It was on the L&A Railroad right of way one mile south of the state line. A grading crew was camping in tents. The L & A tracks lacked about five miles of reaching here at that time but were soon laid. The Pine Woods Lumber Company sent George Harris as general manager and J. F. Giles as secretary and commissary manager to begin building a sawmill. I was present when Mr. Giles pulled the sign, 'Barefoot Station' from the store door and wrote 'Springhill' on the reverse side and nailed it up again. That was a few days before Christmas, 1896.[6]

There was no committee, no political debate, no vote. But the name 'Springhill' was given to the little village where Bodcaw Lumber Company was under construction, and it has remained henceforth. The name has become associated with the kraft paper industry around the world. It is displayed proudly on maps of the nation by athletes, business men, and professional people who graduated from the local high school.

John Murrell.[7] In the winter of 1818, John Murrell, his wife and six children left Carthage, Tennessee, with a few household goods, cooking utensils, a pack horse, two dogs and a rifle. They traveled by barge down the Cumberland, Ohio, Mississippi, and Red Rivers seven years before Louisiana became a state.

Ten families joined the Murrells at Nashville, Arkansas: Wallace, Clark, Ward, Manning, Dyer; Hutson, Robinson, Duty, Dooly and Peterson. Other early settlers already in the area included Sims, Butler, Peters, Rainey, Pipes; Sikes, Alden and Fields. Murrell settled in Claiborne Parish (part of it became Webster Parish in 1871).

Murrell built a two-story house known as Flat Lick Plantation. It had eighteen rooms and two chimneys made of native stone. The house would serve as the first church, first school, first post office, and first courthouse of Claiborne Parish. The first road in Northwest Louisiana became known as Military Road. It was built in 1828 and it passed in front of the Murrell house.

Old Springhill depot west of the railroad tracks.

John Murrell is buried at the Murrell Cemetery on his home place, Flat Lick Plantation. Although he did not live in Springhill, he did live in close proximity to it in Claiborne Parish. It is possible, perhaps likely, that settlers from his family or from other families who came to Claiborne Parish with him visited or settled in nearby Springhill. Nevertheless, he influenced the Northwest Louisiana region by bringing a civilized life style to his new home.

William Buchanan (1849–1923).

Eight-up oxen log team passing old cotton gin around 1914.

J. McKenzie saw gang ca. 1914.

New Springhill depot east of the tracks, ca. 1948.

Chapter 4

Saplings–Youthful City

Pine Woods Lumber Company

Charter. A corporation was legally formed in Lafayette County, Arkansas, in February, 1897, by William Buchanan, W. T. Ferguson, G. E. Harris, Robert Buchanan, J. F. Giles; W. C. Brown, T. A. Brown, M. Northcutt, J. G. Ferguson, and J. A. Buchanan. The purpose of the corporation was to conduct a steam sawmill for the manufacture and sale of lumber. The first article stated, "The corporation shall be styled the Pine Woods Lumber Company, Limited, domiciled at Springhill in the Parish of Webster."

The original charter of Pine Woods Lumber Company was eventually amended May 29, 1914. A resolution was passed to amend the original charter, Section 2, to read:

> Also, in connection therewith, to carry on a general mercantile business, to establish, maintain and operate a supply store; to deal in commrce, buy, sell drygoods, groceries, hardware, and all other kinds of merchandise and materials, including corn, cotton, and country produce; and to prosecute and carry on all matters incidental to in any way properly connected with said business as herein before described.[1]

Fire. The Springhill mill burned in 1912. Buchanan turned his negative experience into a positive one by building a new mill, bigger and better than the former one. He brought in from the Bodcaw Lumber Mill in Stamps, Arkansas, an iron smith named Pompey Flowers. Pompey was a boyhood friend of Dr. Rupert Butler in Buncom Hill. People said he always wore a skull cap which he would tip when meeting a "white brother." His blacksmith shop was located at the south end of Giles Street and east of the L & A Railroad tracks near the mill. He and his assistant did all the ironwork for the new mill.

Day to day operations.[2] The work began when the mill whistle blew at 7:00 A.M. Children in mill towns remember the clanking and rumbling of logging trains pulling out into the pre-dawn darkness carrying men to work in the forest. The payroll office and the commissary remained open to accommodate the loggers who left before dawn and returned after nightfall.

Logger crews were all a part of a series of specialized inter-working teams that were very competitive with one another. The crews had tree fellers or flatheads, saw filers, mule skinners, swampers, scalers and rail gangs. Each one had a specialized job that contributed to the entire process of cutting trees and getting the logs to the mill.

Hazards of work were worse than any other phase of the lumber-producing business. One worker described the men in pitiful terms: "It was killing. I can remember the woodsmen coming by, and I know they were young men. They were bent as badly as I am now. And it was just from work. As a little child I'd think they were old men."

When logs reached the mill, they were handled by workers specializing in their jobs: log pond operators, carriage riders who dogged the logs into position to saw them, saw filers, planer mill workers, sawyers, stackers, and shipping personnel. Noise, sawdust, whistles, and dust were constant companions of the mill workers.

Company housing. Dr. Rupert Butler and his wife Kate moved to Springhill in 1917. Mrs. Butler observed the divisions in the town and wrote, "The town is divided into sections, 'On the Hill'—eleven business houses on the Hill—Mill town or Egypt town, Camp Quarters—Across the Branch—the Front Row."[3]

This confirms the observation by Mrs. O'Bier. She wrote,

> As the village of Springhill developed, it was divided into sections. Front Row, The Hill, Mill Town or Egypt, and the second section, Split Log Quarters and Saw Mill Quarters.
> From the sawmill north was Front Row. The streets ran on each side of the railroad from the mill to the train depot. No business could be on Front Row but Buchanan's.

Since the company owned Front Row or the sawmill area, others who wanted to build any kind of business had to build on "The Hill." When the paper mill was constructed in 1937, there was a severe shortage of housing. Many people lived in tents in the city park, which was called appropriately "Tent City."[4]

Mill town life. The towns where William Buchanan built his seven sawmills were similar in appearance. They were laid out in square grids with shotgun houses designated as "sections" for whites, African-Americans, and company management. Each house faced a street and backed on an alley. Most houses were painted gray. A longtime African-American resident of Springhill called the houses in the black section paper houses.

Sanitary conditions were deplorable. Outhouses were built on flat ground near streams or near a small ditch. Flies were everywhere. Yards were fenced and populated with animals. Houses were close to the mill in order to share electricity with the mill generator and to offer workers easy access to their jobs. Noise, fumes, and debris from the mill became an inescapable part of everyday life. The most prominent and best remembered noises were the company whistles which signaled the beginning and ending of the work day and the noon meal. Also, when the wind blew from the south, cinders would cover everything.

An investigator for the United States Commission on Industrial Relations toured many towns in East Texas and Western Louisiana. He reported, "The land and all the buildings—hotels, houses, churches, schools—are owned and controlled by the company. The people are depressed, shy, and on the surface appear to be satisfied with their lot. There is one organization and no open criticism of the company."

Commissary. The commissary owned by Pine Woods Lumber Company was located at the corner of Giles and Ensey Streets, the present site of Citizens Bank and Trust Company. The author's mother, Flossie Bonner, worked in the drug department of the commissary in the mid-1920s. A fire destroyed the store during WWII.

Closing. Kate Butler wrote in her unique way, "Pinewoods mill was built in 1896. Mr. Harris first superintendent of Pinewoods mill. J. F.

Giles superintendent beginning in 1914. Pinewoods Lumber Company operated 37 years—cut out in 1933."

Frost Lumber Industries

Purchase. In 1936 Frost Lumber Industries, Inc. bought the holdings of Pine Woods Lumber Company and resumed operations of the local mill on May 15, 1936. They bought also the Mobile Bulk Plant which had been operated by T. O. "Top" Machen, whom they retained as agent. The bulk plant was located on the northwest side of the intersection of present day Main Street and Church Street. Mr. R. A. "Buck" Smith was manager of the Frost Lumber Industries in Springhill. Those who knew him described him as "firm but fair" in his management style.

Operation. Frost Lumber Industries, Inc. operated the sawmill from 1936 to 1946. Frost specialized in oak slats for wine vats which were sold to Italy and to Spain. Stealing the oak slats became so prevalent that the company began shipping whole white oak logs overseas.

Sale. Frost Lumber Industries, Inc. operated the sawmill in Springhill until 1946. Anthony Forest Products, Inc. bought the mill and timberland, but not the commissary, from Frost Lumber Industries, Inc. in 1946.

Anthony Forest Products, Inc.

The beginning of Anthony Forest Products, Inc. can be traced back to the grandfather of Clary Anthony. He began the sawmill operations and willed the company to his four sons, one of whom was Frank Anthony, the father of Clary Anthony. As the number of sawmill investments increased, Frank Anthony brought his children into the business to serve in leadership positions. Clary Anthony became Chief Executive Officer of Anthony Forest Products following the term of his brother, Melvin Anthony.

Management. Melvin Anthony was president of the company from the time of purchase in 1946. In addition to his supervision of the mill operations, he lobbied

Governor Faubus of Arkansas to pave Highway 132 from Springhill, Louisiana to Taylor, Arkansas. Mud was so deep in the highway that it required a team of mules to pull cars out of the deep mire. Melvin Anthony was interested in paving the highway to provide an avenue of transportation from the rural areas into the mill area. International Paper Company had a similar goal of getting their workers from rural areas into the paper mill area.[5]

Sales. Immediately following World War II lumber was in great demand. A growing and prosperous nation needed lumber for the housing boom that occurred after the war. Suburbs were being developed rapidly. Lumber was shipped to many parts of the nation from the Springhill mill.

Business expansion. The company expanded into the insurance field by purchasing the Bryan Insurance Agency. It became known as the Anthony-Bryan Insurance Agency. Billy Bryan continued to manage the agency until his death, at which time Wally Hull became the manager.

Closing. The Board of Directors made the decision to close the Springhill sawmill in May 1972. The company had purchased a plywood mill in nearby Plain Dealing, Louisiana. The Board decided there was not enough timber in the area to operate both plants.[6]

Fire. The remaining structures of the sawmill were destroyed by a massive fire in September 1974. All remnants of the sawmill are gone. The mill begun by William Buchanan in 1896 called Bodcaw Lumber Company has now passed into history.

Frank Anthony Park. The acreage where Bodcaw Lumber Company, Pine Woods Lumber Company, Frost Lumber Company, and Anthony Forest Lumber Company mills were located is now cleared and used as a motor home park with facilities for overnight guests. The park is named in honor of the patriarch of the family, "Frank Anthony Park."

International Paper Company[7]

Construction. Pine Woods Lumber Company and Springhill Bank had closed their operations during the Great Depression, leaving Springhill in a difficult situation economically. Then came the break that set the course for Northwest Louisiana to have one of the greatest industrial expansions in history—the decision by Southern Kraft, a division of International Paper Company, to locate one of the largest paper mills in the world in Springhill.

The new paper mill was the catalyst that changed a declining sawmill town into a thriving boomtown. The good times would last forty-two years.

Early one September morning in 1937 construction on the paper mill got underway, plunging the area into a period of activity much like the bustling nineteenth century gold rush in California. Some of the most well-known pulp and paper engineers in the country designed and engineered the construction of the facility from the ground up. Headed by Erling Riis (after whom Lake Erling was named) the list of top engineers includes George Ward, Bill Davis, S. S. McGill, Arthur Foster, Arthur Perkins; H. D. Shope, and Wilburn Lowes.

Operations. Construction had progressed far enough by the summer of 1938 to allow the bleach plant to begin operations on June 20. Six days later the number three machine produced the first bale of bleached, machine-dried pulp. Other construction projects have been accomplished through the years: pulp drying machine (1938), air pollution reduction (1946), number four machine (1952), milk carton paper (1954), seven-thousand acre lake reservoir (1956), and a polyethylene extruder (1967).[8]

Contributions. The city of Springhill provided International Paper Company with a skilled labor force, natural resources, and a good quality of life for the people who lived in the town. These are recognized and commended. However, International Paper Company provided enormous resources to the town of Springhill. Contributions can be enumerated in long reports, but four basic ones from the company should be noted.

First, the mill attracted highly-skilled and well-trained employees in many fields. These talents were used in community service, organizational participation, political life, churches, and social and service activities. The life-style and educational level was enhanced by individuals who moved to Springhill to work at the paper mill.

Second, economic contributions are obvious. The monthly payroll impacted positively every business and institution in the city. Springhill felt the harsh reality of losing 1000 jobs when International Paper Company closed its production operation in 1979.

Third, charities were strongly supported by paper mill employees. The Webster Parish Society for Crippled Children and Adults, North Webster Parish Givers Fund, and the Springhill Unit of the Louisiana Heart Association are a few of the charities that were recipients of funding from International Paper Company employees.

Fourth, additional funds of $25,000 were presented to the North Webster Parish Industrial District for an economic impact study. A 100 acre tract of land was presented to the Industrial District by International Paper Company to use for new businesses that would locate here.

In addition to the specific gifts the paper company offered amenities that provided for a better quality of life. The golf course and Lake Erling are two examples of the company's contributions to the life-style of the city.

Future operations. John Nevin announced the closing of the mill's production division to a group of Springhill businessmen in October 1978. In the last of the five items of his speech Mr. Nevin said,

> International Paper Company has decided to construct a major wood products complex in Springhill. The complex will produce lumber and plywood. It is expected to start up about eighteen months after receipt of the environmental permits. The wood products plant will employ approximately 385 people when it is in full operation.

Business Leadership

Bank. The Bank of Springhill was established on November 4, 1916, in the town of Springhill. The bank was organized because merchants on "The Hill," independent of Pine Woods Lumber Company and William Buchanan, decided they needed a bank they could control. The leadership of the lumber company heard of their plans. They offered the new group of bank owners lots for sale near the sawmill. The Board of Directors of the Springhill Bank were J. F. Giles, Harry H. Crockett, R.A. Smith, R. L. Ensey, G. I. Reynolds; and E. M. Crockett. This board accepted the offer to purchase the lots on Giles and Ensey Streets on the east side of the railroad tracks and the sawmill. This decision created a move of businesses into the bank area that eventually became the downtown area.

Small businesses. A two-story building was constructed on Ensey Street to house the Bank of Springhill. The bank was located on the first floor. Dr. Butler and Dr. Dillon practiced medicine on the second floor.

Other businesses developed along Giles Street: Arnold McLaren's barber shop, post office, Tennyson's Drug Store, Hotel Manuel O'Garte, and the telephone office.

Springhill has been blessed with strong business leadership since its founding: J. F Giles, John Browning, Ed Shultz, Jesse Boucher, Johnny Herrington; Melvin and Clary Anthony, James Branch, Eugene Waters, Johnnie Hill, and plant managers for International Paper Company.

Summary of growth. The village of Barefoot consisted of a small group of farmers who migrated there after 1810. The construction of Bodcaw Lumber Company began a surge of growth between 1896 (date of construction) until 1937 via the lumber industry. The economy was based on the sawmill. In 1937 International Paper Company built one of the world's largest paper mills in Springhill. This dramatically changed the city and initiated a period of rapid growth that would last until 1979 when the company closed the production machines in the mill. Between 1896 and 1979 the youthful city added small businesses, banks, and small industry to the economy. The young sapling was beginning to grow.

Manuel O'Garte Hotel in the 1920s, on Giles Street.

Right: The Bank of Springhill on the day of car giveaway.

Below: Pine Woods Lumber Co. Commissary car giveaway.

First Texaco Bulk Plant, early 1930s: Reuben Haynes, T.O. Machen, Rubin O'Neal.

A special day in Springhill.

City Café, on east side of Main. Riley and Cora Barnard, Irma Andrews behind counter. Bob Barnard seated in front.

Drug Dept. Pine Woods Commissary, Flossie Tapscott Bonner behind counter.

First Ford dealership, at northwest corner of Main and Hwy. 157E.

Springhill Mercantile, John Browning near front.

Machen's Café in 1937; Mrs. Ella Machen behind counter.

Local gas station in 1926.

Ritz Café, at northeast corner of Main and Ensey.

Sawmill engine—last vestige of an earlier era.

Millpond reflects sawmill.

Machinery facilitated the loading of logs in modern times.

Sawmill in operation in 1947.

PART THREE

Rapid Growth–
The Forest Expands
1937–1979

Chapter 5

The Greatest Pine–International Paper Company

In 1937 International Paper Company selected Springhill to be the site of its new Southern Kraft Division paper mill. This decision changed forever the history of the small sawmill town and created an industrial center in Northwest Louisiana. Because it has been the largest employer in Springhill and because it has created the greatest impact on the economy of the city, this entire chapter is devoted to International Paper Company. It is the largest of the pinecones on the tree.

History

International Paper Company. International Paper Company was founded in 1898 when a number of small mills merged into one large company to take advantage of the paper making process. A large newsprint mill was constructed at Three Rivers, Quebec, about 1921. At the same time the Canadian subsidiary was formed, International Paper Company began looking toward the South where pine trees were abundant and had a cycle of fast growth. The purchase of the Bastrop, Louisiana, mill in 1925 marked International's entry into the kraft paper field in the South.[1]

Springhill mill. The Springhill mill was constructed in 1937. Operations began in 1938 in an area which a few months before had been an open field. Springhill was originally built to manufacture bleached pulp for the rayon industry and heavy board or brown paper for corrugated shipping containers. But with subsequent additions and the development of new techniques Springhill shifted production emphasis to kraft container board of bleached kraft paper and board. Within six months of its conversion to these products, the mill was producing fine heavy-weight bleached kraft grades which were soon to become the standard for the industry.

Management

Men of vision. The plant in Springhill stands as a reminder of the pioneering spirit of R. J. Cullen, past president of International, and Erling Riis, past president of International and general manager of its Southern Kraft Division.

Richard J. Cullen possibly did more than any other man to further the development of the paper industry in the South. He was one of the first to open up the vast resources of Southern woodlands for the raw material for making kraft paper. As a young mechanical engineer he designed and built the Bastrop International mill and the Louisiana Pulp and Paper mill in Bastrop. He also directed construction of the mills at Camden, Arkansas, Moss Point, Mississippi, and Mobile, Alabama. He became manager of all International Paper Company mills in 1935. In 1936 R. J. Cullen was elected President of International Paper Company.[2]

Erling Riis was born in Norway. He began his career with International Paper Company in 1920 in Bastrop, Louisiana. He was chief engineer on the construction job of several mills, including Springhill. In 1946 he became Vice-President of International Paper Company. Riis was considered one of the foremost engineers in the paper industry. He planned and developed some of the most modern paper mills, originated many labor-saving devices, processes, and products, and is considered an outstanding authority in efficient construction and operation of pulp and paper mills. Two other men who were influential leaders in International Paper Company were John H. Hinman and J. H. Friend.

John Hinman became President of International Paper Company after R. J. Cullen was elected Chairman of the Corporation. Mr. Hinman was a strong believer that the present security and future growth of the U. S. pulp and paper industry depended upon the widespread adoption of scientific woodland management practices. He led International Paper Company to carry on intensive programs of woodland conservation and tree farming.[3]

J. H. Friend was Vice-President and General Manager of five International Paper Company mills. Later he was promoted to Director of International Paper Company on May 10, 1939, and was President of the Southern Kraft Timberlands Corporation. He was known as a progressive manager who kept all mills apprised of the latest developments in the industry and maintained up-to-date repair and modernization programs. Friend was very active in fire control, insect and disease eradication in timberlands, and in stream improvement of water pollution problems.[4]

Construction and Operations

In September 1937 construction on the paper making operation got underway, plunging the area into a period of activity. Before the introduction of International Paper Company to the area the population of the two towns of Springhill and Clifford (Cullen) was about 850 people. Forty years later the area boasted of a population of 8,354, a growth of 7,500 people.

Land. Mr. Pat Krouse, a resident of Springhill who retired from the company in 1961, and Mr. R. A. Smith, Manager of Frost Lumber Industries, assisted International Paper Company in obtaining options on the land for the mill and timberlands in the area.

Some of the land was sold to International Paper Company by Mr. Henry Hood and Mr. Bud Coyle. Henry Hood was fishing when two well-dressed men approached him in an effort to purchase his land for the new mill. He had homesteaded the land and lived on it since he came from North Carolina. He sold the land which became a part of the acreage used by International Paper Company upon which to build the mill. Mr. Bud Coyle was a successful farmer in the Clifford area. He sold acreage to International also.

Engineers. Some of the most well-known pulp and paper engineers in the country designed and engineered construction of the facility from ground up. The team was headed by Erling Riis.

Projects. Construction projects over a period of time were described by the editor of The Springhill Digester, Harry Brewton. He wrote the following chronology:

> The original plans called for the installation of a two-machine mill with supporting pulping, power, and shipping facilities for the production of 600 tons per day. One machine was to produce corrugating medium linerboard. Soon after the foundation had been set for the facilities, company officials decided to build an easy-bleach pulp mill, a pulp drying machine for the production of 150 tons per day of bleached, machine-dried pulp. The first major attempt to conserve saltcake and reduce air pollution was made in 1946–1947, with the installation of electrostatic precipitators on the six chemical recovery ducts, and some increase in evaporator capacity was added. In December 1952 the number four machine was started up on cylinder board. The second-hand machine came to Springhill from Hartford City, Indiana. The demand for paper milk cartons began to surpass all expectations and in 1954 it was decided to produce bleach grades on the number three machine. This adjustment in operations resulted in the addition of a lime kiln, two digesters, a set of brown stock washers in the kraft pulp mill, additional barking drums, and the number six power boiler. During this period the shortage of water became critical due to lack of rain and the increased demand for its use. A seven-thousand acre reservoir was built about ten miles north of the mill in Arkansas, and was dedicated "Lake Erling" in 1956 in honor of the late Erling Riis who headed the construction of the mill. Emphasis was placed on improving the product cost, customer service, maximizing the use of resources, replacing old ideas with new improved methods and equipment, and improving the mill's quality of air and water emissions during the 1960s. The polyethylene extruder for coating such grades as cup stock and linerboard were installed in 1967. It was during the early and mid-1960s that modernization of the power/recovery plant continued with the installation of a 500-ton and a 700-ton chemical recovery unit, both equipped with electrostatic precipitators to further improve the quality of emissions. During the last decade of the paper mill's operations, the company invested almost $35 million to comply with environmental requirements and improve the mill's production efficiency.[5]

June 20, 1938. Construction had progressed far enough by the summer of 1938 to allow the bleach plant to begin operations on June 20th. Six days later the number three machine produced the first bale of bleached, machine-dried pulp.

December, 1952. During this year the demand for milk cartons began to surpass all expectations. In 1954 management decided to produce bleach grades. The shortage of water caused the company to build a seven thousand acre reservoir called Lake Erling to provide water when needed.

1967. The polyethylene extruder for coating grades of paper for cup stock or linerboards were installed.

Closing

Date. Mr. John Nevin, Manager, Manufacturing of International Paper Company's White Paper Group, met with Springhill business leaders on Tuesday afternoon, October 17, 1978, and made the following five-point announcement:[6]

1. I am meeting with you today to announce that International Paper Company has decided to terminate production of liner board and uncoated bleached board at Springhill early next year. This means that nos. 1 and 2 machines and supporting operations will be shut down at that time. Arizona Chemical, the Container Plant, Woodlands Office, and an interior utility source will continue to operate. Sheet finishing will continue to operate if we can make the necessary changes to make it a competitive stand-alone operation.
2. With reference to the remaining facilities, the Container Plant and Arizona Chemical will be equipped to operate as independent facilities. Package boilers will be installed at the Arizona Chemical and Container plants to provide steam requirements. We will continue to operate a power boiler on an interim basis until installation of the package boiler is completed. Wells will be utilized to furnish water requirements. Fencing will be provided to separate and enclose the facilities as separate identities.
3. Lake Erling and Williams will remain in place. Waste water will be drained and secured except for SB-1 which will continue to be utilized by the remaining facilities. The Woodlands office will continue to function and receive truck and wood at the Springhill site.
4. Abandoned equipment will be relocated to other IP facilities if needed, if not needed, will be sold.
5. Approximately 1000 people will be affected by this decision. We will be meeting with you to discuss specific positions affected and appropriate benefits. These meetings will be in approximately 30 days as scheduled by local management. Employees interested in finding employment at other IP mills can make application through the personnel office which will have knowledge of permanent vacancies in other IP operations in the South.
6. IP has decided to construct a major wood products complex at Springhill. The complex will produce lumber and plywood. It is expected to be ready to start up about 18 months after receipt of environmental permits. The wood products complex will employ about 385 when it is in full operation.

Production ceased February, 1979. The Mayor, Johnny Herrington, said it was like a funeral procession that night. Lines of automobiles passed by the plant where there was no activity. He also said,

> I am 48 years old ... I was born in Bastrop, Louisiana, a paper mill town ... that's all I know, that's all I've known. Although I knew that the mill was going to cease operations, I could not make myself believe it. This (the possibility of the mill closing) is basically why I became involved in the mayor's race in February ... We know that there's a lot of work ahead for all of us, and hard times, but we've all been through hard times before. Although we don't like to do it, we can still do it again. We are confident that through our work and efforts that we will regain our present economic standards, and at the same time be more diversified. This will assure us that what has happened here recently will never happen again.[7]

Tylon Blanton, President of the Webster Parish Police Jury, said,

> It's the most unfortunate thing that's ever occurred in this area, and it affects the life of everyone in the entire area in one way or another. Its going to take time to recover from it, and its going to take a lot of combined effort, but we have a lot to offer industry in this area ... the type of people we have ... climate ... our natural resources. It's just a matter of time until the void will be filled.

Dorothy Smith, Webster Parish School Board member, said,

> It's sad, but we'll survive. I don't think we need to sit here and wring our hands and cry. We'll make it ... we always have ... if everybody will get out and work for it ... I mean everybody in Springhill ... We've got too many things going for us not to make it.[8]

These remarks and similar ones made by dozens of citizens, indicate two things; first, Springhill was living through a grief experience, and, second, progressive citizens were planning to work hard to restore the economy to its previous level.

Four viewpoints. Reasons for closing International Paper Company's Springhill mill are multiple, but four viewpoints have prevailed: cost factors, management re-organization, internal politics, and external politics.

The Fall, 1978, issue of *The Digester* thoroughly summarized the high cost problems of the mill that decreased profits. Page Williamson, Director of Business Operations of International Paper Company, confirmed the cost problems in a personal interview.[9] Mr. Williamson also summarized two other viewpoints on closing the mill—management re-organization and internal politics.

The first viewpoint on closing the mill examined cost factors and mill obsolescence. The scenario goes something like this: The post war expansion that added machines No. 3 and 4 created two mills in one area. The machines were placed on the other side of the mill away from machines No. 1 and 2. This distance created enormous cost problems necessary to operate the paper machines. Profits were small on brown side paper because costs were up and sales prices were low. Furthermore, the Georgetown mill and the Texarkana mill were competitors with the Springhill mill for profits. The question was asked by management,

> Do we want to spend $500 to $600 million in Springhill to bring the mill up-to-date with new equipment or do we want to spend the same money to build a new mill in Mansfield with current technology and less manpower to operate it?

The decision was made to build a new mill rather than remodel an old mill.

Other cost factors entered into the equation. Federal environmental laws that relate to clean water and clean air are constantly more demanding on companies to clean the air and purify the water. This is an expensive operation. Also, management wanted the union to go to multi-craft jobs and write it into the forthcoming contract. There was resistance to this idea. Cost for labor was more for the old contract jobs. Water supply was from deep wells and limited. This, too, was expensive.

The second viewpoint on closing the mill was centered on management re-organization. When the author pastored a church in Huntsville, Texas, one of the deacons who was a forester told him about the hiring of J. Stanford Smith from General Electric to be President of International Paper Company. Management re-organization was an idea whose time had come. Stanford Smith led the company through change from manufacturing control of production to business management control of production. The idea was to bring cost factors down and earn more profits. This change occurred in 1976, two years before the announcement that the Springhill mill was being closed by IP. The profit motive decision centered around white side versus brown side paper. White side was more profitable, but the manufacturing group wanted more brown side paper produced. It cost more in an old mill and made less profit. The new management style placed business managers in control of the production decisions which were based on profits. White side paper was emphasized. There was a goal to develop a world class production mill that caused management to make a decision to build the Mansfield mill.

The third viewpoint on closing the mill was based on internal politics. A difference of opinion about the type of paper produced was expressed in many forms. Brown side paper was being produced by the tons, but it was less profitable and cost more in the Springhill mill. Other factors entered into the political discussions. The water problem in Springhill was not good. There was no interstate near the mill. Don Brennan in the senior management level of International Paper Company was committed to a world class mill rather than remodeling an old one. Other mills like the ones in Texarkana and Georgetown were producing more paper at a cheaper cost.

During this same time frame International Paper Company sold C.I.P., Canadian International Paper. Millions of dollars were now available to build a new mill in Mansfield. Resources were available, environmental problems were prevalent, and profits for brown side paper in Springhill were lower. John Nevin, Page Williamson, and Cecil Bailey were in a group to cut cost and produce white side paper, but the political decision was made by management to close the brown side production in Springhill and move it to a new mill in Mansfield.

The fourth viewpoint on closing the Springhill mill is based on the popular citizen's belief that the decision to terminate production at the Springhill mill was an external political one. There is no documentation for this viewpoint. It is based on speculation. The scenario goes like this as it was expressed to the author by various citizens. The paper mill effluent in Bodcau creates problems. Bossier City wants a clean water supply from Bodcau. Real estate values will go up with clean water. Environmental problems are severe. Therefore, political influence from outside Springhill created enough pressure to build the Mansfield mill and close the one in Springhill.

There is no evidence within International Paper Company nor in any documentation that this occurred. It is, therefore, left in the realm of personal opinion of individuals.

Closing the Springhill plant should not have come as a surprise to the citizens. Warning signals about high cost factors and environmental problems had been posted as early as 1975. *The Digester*, the quarterly publication produced for employees of the mill, contained articles about environmental problems in the Spring 1975, Spring 1976, and Fall 1977. The Fall 1978 issue carried an article on the shutdown of machine No. 3 which reduced employment about 450 persons.

The announcement of the shutdown of No. 3 machine had been made April 17, 1978. Between that date and October 10, 1978, when the announcement was made to close the mill, manager Cecil Bailey led the employees to make the facility cost competitive. He announced to the employees at the conclusion of the cost cutting program,

> We said from the outset that the mill was in tough straits, and all employees have done their best to help us make a go of it. We couldn't have done any better because we did our best. Our costs are simply too high, when compared to the rest of the company's production capacity, and our competitors' as well.

John Nevin said,

> There have been wholesale cost increases in all areas of mill production in labor, fuels, construction, chemicals, and other materials, causing the Springhill facility to become one of the least efficient mills in the company.

Contributions

Before the mill closed in February 1979, International Paper Company donated two tracts of land to Springhill. In March 1978, an 82-acre tract was donated to the city for expanding the airport. In the comments following the presentation of the deed to the property, mill manager Cecil Bailey remarked,

> The company is actually interested in the welfare and future of the citizens of Springhill, Cullen, and Webster Parish. The donation of this land to the city provides a key element to the area's efforts to promote industrial growth for local businesses and citizens of the area.

In August, 1978, a 100-acre tract of land located just south of the Springhill mill site was donated to the newly established north Webster Parish Industrial District by the company in an effort to assist the local government officials in promoting industrial growth for the area.

John M. Nevin, Manager of Manufacturing for the White Paper Group, said,

> IP has appreciated the unselfish support given by the Springhill-Cullen community. We are pleased to demonstrate this appreciation by donating this land, thereby, opening up opportunities for industrial development here.[10]

1949 photo of I. Y. East, I.P.C. manager.

International Paper Company, 1948, with entrance to Pine Hill Subdivision in lower right.

Springhill Mill about 1949.

International Paper Company barbecue pit and pavilion, Emmett Hull and E. W. Brown, early 1950s.

Local citizens observed L.S.U. experimental gardens using black water irrigation.

Rantz Mason, Sam Williamson Sr., and E. S. Krouse taking black water samples.

I.P.C., February 1979, not long before mill closing.

L.S.U. School of Forestry (engraved on cross section of log by lumberjack's boot) with Ray Morgan and John Tullett ca. 1951.

International Paper Company, February 1979.

Southern entrance welcome sign to Springhill in 1948.

Chapter 6

Great Pines–Economic Growth

Springhill experienced rapid economic growth between 1937 and 1979 (construction and closing of International Paper Company). Other plants added to the paper industry to create economic growth in the bustling Webster Parish city. Arizona Chemical, Stauffer Chemical, Consolidated Chemical Industries of American Cyanide, Nations Brothers Packing Company, Springhill Lumber Company, Sanitary Dairy, C. & S. Ready Mix, Crusader Drilling Service, E. S. Sikes Pipeline Contractor, and E. B. Smith and Sons were some of the other companies built in Springhill.

One business journal stated, "There is no sign of easing off at the present time. Springhill continues to grow, and good industrial sites are readily available—sites that are well-drained and accessible to transportation connections."[1]

The economic photograph of Springhill has not always been so colorful, nor would it remain that way. Dark days before that time had been experienced by the city.

Over the decades Springhill changed from an agrarian city to an industrial city. Since 1896 the economy was based on the sawmill. In 1938 International Paper Company began production of kraft paper which initiated an era of unprecedented growth. Then, the production part of the paper mill closed in 1979. The sawmill had closed in 1972. Once again economic times were difficult. Between the construction of the paper mill in 1937 and the closing of production in 1979 the economy mushroomed.

Through the years Springhill experienced four eras of economic growth. The first was an agriculture economy.

Agricultural Economy 1810–1896

Beginning about 1810 early settlers migrated from Middle Tennessee and from the southeastern part of the United States into Northwest Louisiana. They made their living by farming, growing their vegetables, meat, poultry, hogs, cattle, cotton and corn. Markets for cotton and corn were too difficult to reach, so the occupation was a poor one.

Sawmill Economy 1896–1937

The agricultural situation of the poor farmer changed in 1896 when Bodcaw Lumber Company was built by William Buchanan in the little village called Barefoot. Soon the sawmill employed large numbers of the farmers.

The name of Bodcaw Lumber Company was changed to Pine Woods Lumber Company. It manufactured lumber until it closed in 1933 during the Great Depression. For two years the economic situation was difficult. Only farming remained once again. In 1936 Frost Lumber Industries bought the Pine Woods Lumber Company and began to manufacture lumber once again. The economy became stronger.

Paper Mill Economy 1937–1979

The real impact on growth and development of Springhill came in 1937 when International Paper Company announced they would build a kraft mill in the city. It began operations in 1938.

Since the mill began operations in 1938 the growth and prosperity of the city has been linked with the growth and prosperity of the paper mill. In 1963 retail sales of small businesses reached $9,000,000 annually. The Kansas City Railroad maintained its lines on the L & A Railroad rails between Hope, Arkansas, and Shreveport, Louisiana, through Springhill. It had the fifth largest freight car loading and unloading volume record of all Louisiana cities.[2]

The discovery of oil and gas in the area coupled with gravel and clay resources added to the good economic times in Springhill. The city became a trade center for much of North Louisiana and South Arkansas.

Diversified Economy 1979–2002

The closing of the manufacturing division of the paper mill in 1979 was a crushing blow to the economy and a wake-up call to the city's leadership. No longer could Springhill depend upon one major industry. Diversification of business and industry was the answer.

The North Webster Parish Industrial District was organized in July 1978 to encourage small industry to move into the area. The North Park section of the District had eight businesses operating in 2000: Office Dimensions, Trane Company, Soil Products, Electric Melting Services Southwest, American 3CI, Morgan Chop Saw, Ironworks Grill, and 3-D Contractors. The South Park section has two businesses: Springhill Pallet Company and Reliant Energy Pipes.

Over 300 people are employed in the industries. The North Webster Parish Industrial District in 2000 A.D. is led by Eugene Eason, Chairman of the Board.

Other Board Members are Barry Slack, Vice-Chairman; Johnnie Hill; Howard Beaty; Charles Jackson; and Emmitte Salim. The staff includes Jean Ervin, Administrative Assistant; Bobby Slack, Water Superintendent; Rex Bryan, Economic Development Consultant; and John Slattery, Jr., Attorney.

Banking Business

As a rule the important role played by the bank is not fully appreciated by the general public. A majority look upon it as a place of safekeeping for their money and a source for small loans. Banks do provide these services, but they also contribute to the success of many small business enterprises. Banking in Springhill has a long chronology dating back to 1916.[3]

Bank of Springhill. The Bank of Springhill was established November 4, 1916. Shareholders were James F. Giles, Harry Crockett, R.A. Smith, R. L. Ensey, G. I. Reynolds, and E.M. Crockett.[4] A two-story brick building was constructed on the corner of Giles and Ensey Streets. The new Bank of Springhill was located in that building on the first floor (across the street from today's Citizens Bank). Murray Tennyson later placed his drug store in the southwest corner of the bank building. The offices of Doctors Butler, Dillon, Young and Attorney Choate were located on the second floor.

Commercial Bank and Trust Company. This bank had a brief life during the Great Depression years. It was organized in 1933 and closed its operation April 10, 1934.

Minden Bank and Trust Company Depository. Since Springhill needed some type of banking service, a Depository was established in 1934. Minden Bank and Trust Company turned the Depository over to People's Bank and Trust Company in January, 1935.

People's Bank and Trust Company. On January 1, 1938, People's Bank and Trust converted the Depository to a Branch Bank in order to serve a fast growing community with a new paper mill. In 1943 the Branch Bank was converted back to a Depository.

Springhill Bank and Trust Company. The Depository was inadequate for the needs of a growing community, so progressive business men formulated plans for a locally owned and operated bank. They obtained a charter from the Deposit Insurance Corporation and opened the Springhill Bank and Trust Company for business in April, 1943.[5] Officers of the bank who developed it from infancy were J. B. Slack, President; J. M. Browning, Vice-President; and D. G. Tyler, Vice-President and Cashier. There were eight Directors on the Board: J. A. Branch, M. T. Browning, Ed Shultz, I. B. Slack, D. G. Tyler, B. L. Slack, J. M. Browning, and Dr. Rupert Butler. Other presidents of First National Bank were Ed Shultz, Eugene Waters, and Ray Huddleston.

In June, 1998, Springhill Bank and Trust Company was sold to Regions Bank. No longer was it a locally-owned bank. However, a local resident, Ray Huddleston, is the President.

Citizens Bank and Trust Company. The bank was created by Charles McConnell and Waymon Oden. It was organized and opened for business on July 1, 1955. When the author asked Charles McConnell about the reason for the development of a new bank, he responded, "We felt like we needed another bank in town for the competition."[6] Several competitive reasons were summarized in that statement.

The Citizens Bank board was composed of the following persons: Charles McConnell, Chairman; Waymon Oden, President; Melvin Anthony, Vice-President; Paul Offutt, Vice-Chairman and Cashier; O. M. Slack, D. L. Booth, L. Manuel, Dr. W. C. Gray, Dr. J. E. Rutledge, C. N. Payne, W. J. Cassells, and G. E. Steed.

The honor of being the first depositor went to M. T. Browning, retired businessman of Springhill. He presented his bucketfull of $20 bills at the teller's window for deposit immediately after the bank opened for business.

Clary Anthony, former C. E. O. of Anthony Forest Products, Inc. sawmill, has been Chairman of the Board since 1985. He has led the bank to expand its services to two other towns in the area.

Small Businesses

Numerous businesses prospered in the paper mill economy. An article in *Newsfolder*, July 1952, featured Springhill. The author of that article wrote, "The activity around Springhill is a good general indication of the prosperity of the town. Up-to-date trends are reflected more strongly in the business section, where modern store fronts and the latest merchandising displays are the rule rather than the exception."[7]

During the summer months each merchant closes his shop promptly at noon on Tuesday. The afternoon he can devote to fishing, to a siesta, or to whatever strikes his fancy.

Merchants in the Springhill business district in 1952 included, but was not limited to the following: Springhill Telephone Company, Bryan's Shoe Store, NZ Cash Store, Robinson Drug Store, Branch Brothers Motor Company, Western Auto Associate Store, Princess Café, J. N. Bond Drug Store, 505 Service Station, Willie Mack Department Store, Pickett and Sons, Boucher and Slack Insurance, Curtis Brothers Grocery, Magnolia Petroleum Company, C. C. Harper's Texaco Consignee, Reynold's Store, Pick and Pay Grocery, Tennyson Drug Store, G. F. Wacker Store, Palace Radio Shop, and M. T. Browning General Merchandise.[8]

Oil and Gas Exploration

Deep beneath Louisiana is a gigantic natural factory that has been operating for millions of years, churning out its black liquid product, crude oil, and a lot of natural gas. The crude oil factory was built by the natural geological forces peculiar to Louisiana within about the past 65 million years.[9]

Caddo Parish oil. The first indication that there might be oil or gas in Caddo Parish or in Northwest Louisiana, appeared in 1870 when a deep well was drilled to supply water for an ice factory in Shreveport.[10]

Operating as the Savage Brothers and Monical, the first well in Caddo Parish of which there is an authentic record was drilled on the Caddo Lake Oil and Pipe Line

Company lease in Section 1, Twp. 20, Range 16, which began rigging up in May, 1904, near what is now Oil City—then merely Annanias flag station on the Kansas City Southern. It was drilled deeper and reported as a gasser January 1906.[11]

Haynesville oil. Excitement and "boom days" that followed the discovery of Homer oil and then the Haynesville pools have been duplicated only by Rodessa. It was on March 29, 1921, after two previous failures, that the Haynesville oil field was assured, with the blowing into production of the Smitherman et al Taylor No. 2, Section 14, Twp. 23, Range 8, from 2855 feet, making an initial production of 5000 barrels.[12]

Shongaloo oil. The discovery well of the Shongaloo gas area of Webster Parish was completed in October, 1921, by Portland Syndicate and J. K. Wadley at Munn No 1, Section 1, Twp. 22, Range 10, from 1662 feet, making an estimated 40 million feet of gas.[13]

Homer oil. The Homer field was discovered in 1918. This was one of the earliest discoveries in the area. Haynesville, Cotton Valley, Shongaloo and Springhill fields followed Homer.

Springhill Oil

Dewey Williams, James Branch, and Eugene Waters summarized the oil activity in the Springhill area for the author. In an interview with Dewey Williams, he identified the following chronology for drilling and discovering oil in Springhill.[14]

Old Carterville field. The old Carterville field on the bank of Bodcau Bayou was discovered in the 1930s.

First Springhill well. In 1951 the first well to reach oil within the city limits in Springhill was drilled near the Arkansas—Louisiana state line north of the Springhill Medical Center Hospital.

Dorcheat to Shongaloo. In 1955 oil was discovered in Shongaloo east of Dorcheat Bayou and north of Highway 157. Sixty-six wells were eventually drilled.

Timothy Road. Ten wells were active 1955–1957 on Timothy Road. Gas was also discovered in that area.

New Carterville field. Between 1960 and 1970 the new Carterville Pettit Lime field was productive. The field was located between the S-curve on the Springhill to Plain Dealing highway and Bodcam Bayou.

Walker's Creek to Spring Branch. In the early 1980s there were 40 to 60 wells drilled to a depth of 11,000 feet between Walker's Creek, Arkansas, and Spring Branch, Arkansas, by the Penzoil Company. Many farmers leased their land and received large royalty checks from the production.

Red River to Haynesville. One of the most productive fields was active 1980—1990 from the Red River near Bradley, Arkansas, to Haynesville, Louisiana. It followed a fault line along the Arkansas state line. Wells were drilled to 11,000 feet by Crystal Oil and directed by Bob Roberts. Wells were drilled on Machen Drive and near the Rodeo Arena in Springhill.

International Paper Company ponds. In 1984–1985 there were 70 wells drilled on the IPC mill pond near Old Minden Road.

Total Area

Springhill in center. When one looks at a map of the oil fields in Northwest Louisiana, it becomes clear that Springhill is located in the center of the oil pools. Oil activity goes north to Walker's Creek and Taylor, south to Cotton Valley, east to Haynesville, and west to Bradley and the Red River. This places Springhill in the strategic center of oil activity in upper Northwest Louisiana.

Summary. The third greatest economic factor in the development of Northwest Louisiana, after river and rail transportation, and the one that has brought the greatest wealth into the region, has been the development of its mineral resources. Although the pine tree has been the basis of the economy in Springhill, oil and gas have been extremely important.

The Lumberjack Cafe, at southwest corner of 4th St. N.W. and Butler.

Pickett's Store in 1948 on Butler St. at 4th St. N.W.; forerunner of the international Pickett Enterprises.

Barney Dennis with horses Buck and Dick pulling cars out of muddy Taylor Highway in 1948.

Arizona Chemical Company, painted in 1994 by Jimmy Harlon Driskill and hanging at City Hall.

Springhill Telephone Co. office faced south on East Church St.

State Theater on west side of Main.

Nations Bros. Packing Co., 1948, on Plain Dealing Hwy.

Pulp wood train wreck about 1947.

Webster Theater on east side of Main.

Rancho Drive-In Theater about 1948, located at current Wal-Mart site.

Boucher Drug on west side of north Main, north of Spring Theater site.

Springhill Bank and Trust, Boucher and Slack Insurance on Main, 1948.

Sanitary Dairy ran milk routes in Springhill.

East side of north Main business district.

Looking south from Boucher Drug toward Mobil bulk plant and Tennyson Drug.

Citizens's Bank and Trust in earlier years.

Spring Theater under construction in 1948 on west side of Main.

Norton Funeral Home facing north on Plain Dealing highway on current First Baptist Church housing property.

Remodeled Springhill Bank and Trust.

Springhill Motor Company sold General Motors vehicles on Main St.

Chapter 7

Great Pines–Educational Growth

In a midsummer conference in 1967 the school leadership of Webster Parish commented on the fact that a generation ago the local school system was recognized as being one of the "lighthouse" school systems in the nation. The school board and superintendent observed that this relative position had been lost, and expressed a desire to restore the schools to a level of recognized quality.[1]

Peabody Survey Report

Recommendations. George Peabody College for Teachers was requested to conduct a comprehensive survey of the school system and to present the recommendations for a long range program of school improvement. One-hundred recommendations were submitted in seven categories: organization and administration, instructional personnel, elementary education, secondary education, physical plant, pupil transportation, and finance and school business management.[2]

Observations. The survey team from Peabody College made two observations: "First, The conclusion is inescapable that Webster Parish must change its schools if it wants to improve them—not change to be different, but change to be better." The other observation was very positive: "Second, The future is bright! Webster Parish can have the kind of educational system and quality of life its citizens want."[3] School boards and administrations have followed this advice and used these recommendations to improve Webster Parish schools by adding programs to the curriculum to encourage learning.

Budget

Funds to improve the operation of Webster Parish schools and funds for Capital

Outlay to build new buildings have increased annually as school boards have implemented the Peabody recommendations.

The General Operating Budget for 2000–2001 is $32,493,392 to operate twenty-one schools with 7,764 students. An additional Special Funds Budget is $12,347,682. This money is used for special operational needs within the system. Capital Outlay and Debt Service budgets are $699,414 for Capital Outlay and $551,680 for Debt Service. Final payment for indebtedness is scheduled for March 1, 2018.[4]

Education in Louisiana

Private schools. Early schools in Louisiana were private church schools, private lay schools, and tutors in wealthy homes. Church schools were confined largely to the Roman Catholic Church while private lay schools were funded and operated by teachers who secured their compensation largely by charging each child a small monthly tuition. In the early days tutors instructed as many children as churches and private academies could afford.

General School Act. There was an evolvement of schools from the Civil War and Reconstruction years (1860–1900) to the turn of the century. During that time very little money reached the children. However, a major step occurred in 1877 when the General School Act created a State Board of Education. The Board began a series of responses to the needs of state schools. In 1912 the Burke Act was passed to provide standards and stabilization to the school districts. Salaries, certifications, and budgets were given form and content. In 1916 the Effective Act was passed by the legislature requiring 140 days attendance at school.

T.H. Harris, State Superintendent of Public Education for Louisiana, wrote, "The story of education outlines the successful efforts of an earnest people to provide for instruction of their children from 1803–1923."[5] The people of the state are profoundly interested in the education of their children.

Education in Springhill

Buildings. The first school of record was in 1895 at Rock Hill on the Plain Dealing Highway or on the Mill Pond Road at Boucher Bridge. It was a one-room cabin that housed an enrollment of 28 students. The teacher was W. A. Miller.[6]

Buildings were constructed from 1900–1957. A series of new school buildings took shape during these years. In 1900 a two-story frame building was constructed. Pine Woods Lumber Company furnished the materials and labor. This was a multi-use building. The top floor was used by the Masonic Lodge for their meetings. The bottom floor was used for town hall meetings, church meetings, and a public school.

The next building for the school system was bought from Pine Woods Lumber Company in 1907 for $1,350. Mr. Pope and Mrs. Mothershed were the teachers.

As the public school grew and the years passed, other construction projects occurred. Land south of the Methodist Church was purchased and a building constructed in 1920 for an elementary school.

In 1927 the old high school (now the Springhill Junior High School) was built to accommodate the growing number of students in the system. The front of the building was once covered with ivy which inspired the writer of the Springhill High School Alma Mater to include the words, "memories of thy ivy walls" in the school song.

With the arrival of the paper mill, more people moved into Springhill causing a rapid increase in enrollment between 1937–1948. More than fourteen-hundred students and forty-five teachers were added to the schools. The elementary building and the gymnasium (now part of Springhill Junior High School campus) were constructed.

The first of three units of the new high school was completed in 1949. A $400,000 bond issue was approved by tax payers to construct the second section which included an auditorium, cafeteria, additional classrooms, music department, and an industrial arts shop.

An elementary school building was constructed in 1957. It was named Browning Elementary School in honor of M. T. Browning, a native of Springhill, who served as President of the Webster Parish School Board 1945–1950.

Consolidation and accreditation. Five schools closed and consolidated their students with Springhill schools: Rocky Hill in 1907, Taylor Academy in 1911, Timothy School in 1923, Haynes District in 1927, and Blocker's Chapel in 1928. Most often consolidation was due to financial difficulties. Each time consolidation occurred it caused the Springhill schools to grow in enrollment and created the need for more teachers. The curriculum was enlarged, new buildings were constructed, and transportation to school was provided—first by mules and wagons and later by buses.

Springhill High School was accredited in 1927. In that same year the State Athletic Association accepted Springhill into its membership. This event gave a boost to the athletic program of the school which continues to this day.

Twelve year plan. Two major changes occurred in 1944 during the war years. Springhill High School joined the other Webster Parish schools in a conversion from the eleven year plan to the twelve year plan. Since that time students have been required to attend public schools for twelve years in order to complete the curriculum and to be graduated.

Compulsory attendance law. That same year, 1944, the Compulsory Attendance Law was passed. The law eliminated the habit of some students to attend classes on an irregular basis. If students do not attend classes on a regular basis today, Mr. Bennie Raborn, the truant officer, pays the family a visit to discuss the problem.

Curriculum changes. Superintendent Noles explained some of the curriculum changes in recent years. Webster Parish Schools now have the following emphases in their curriculum: Louisiana Education Assessment Program (LEAP) that tests students on their academic achievement, the Graduate Exit Exam (GEE) that must be passed in order to be graduated from high school, the Accelerated Reader Program, the Comprehensive Phonics Program for Kindergarten classes, the Regular Core Program for all grade levels, the addition of technology into the classroom, and the Character Counts Program designed to improve behavior and instill values.

Philosophy. All of the programs have been added in order to provide quality education for all students in all grade levels and to continue the drive toward developing schools with a "level of recognized quality." The official mission statement of Springhill High School defines its philosophy in its efforts to attain these goals:

> The mission of Springhill High School is to attempt to aid all students in the process of developing attitudes,interests, and goals which will allow them to become problem solvers in their schools and community. Good citizenship and intellectual capacity are fostered through a strong academic program and a well-rounded program of extra-curricular activities and sports. We also attempt to instill in the students the desire and need to continue the learning process throughout the rest of their lives.

Hall of Fame. In 1997 alumni were invited to return to Springhill in order to recognize individuals of Springhill High School who have made a significant contributions in sports or in their careers. Dr. Woodrow Turner is President of the Alumni Association and director of the Hall of Fame events. Jesse Boucher started the program.

The first group of inductees in 1997 were: W. D. Baucum, Jesse Boucher, John David Crow, Michael Haynes, Louise Wardlaw Lewis, Donald G. Mack, Sr., M.D., William Alexander Miller, William Byrd Smith, James Turner, and Samuel Ruthven Williamson.

The second group of inductees in 2000 were: Gary Boucher, Jeanne Mack Gilley, B. J. Hodge, Arthur Logan Miller, Jim Montgomery, Robert Charles Smith, Ted Souter, and Joe Stampley.

The inductees for the 1997 Lettermen's/Letterwomen's "S" club were Ellis Cooper, M.D., John David Crow, Coach Travis Farrar, Earl Haynes, Johnny Herrington, Frances McGowen Hilburn, John M. Montgomery, Cherokee Rhone, Jack Rogers, Jerry Wayne Sessions, M.D., and James Turner.

The inductees for the 2000 Lettermen's/ Letterwomen's "S" club were: Gus Boucher, Sherry Boucher-Lytle, Ophelia Carroll, Leonard Coyle, Gerry Colvin Dumas, Ruth Malone Jones, Raydell Smith, Zolen Stiles, and Charles E. Tyler.

The value of this activity has been the recognition of individuals from Springhill High School for their life-time achievements, but the renewal value of recognizing characteristics of accomplished people has focused attention on qualities that have made Springhill a growing, progressive town through the years; qualities of hard work, strong values, academic achievement, and constructive use of personal skills.

African-American Education

Civil War influence. During the years following the Civil War there was little progress toward education of African-American children. As slaves the African-Americans were not allowed to learn to read. Slave owners kept them illiterate in order to control them.

Eventually small groups of African-Americans would meet under brush arbors for worship. Education was included in the meeting. During this time crude cabins were constructed for the children's education. Most of them were wooden frame buildings which were poorly equipped for teaching.

Buchanan influence. African-American education in Springhill began when William Buchanan built the sawmill in 1896. He brought with him his own African-American workers for the mill. They were well-trained, loyal, and hard working.

Buchanan's policy was to build the mill, houses for employees, and a commissary. Often he added buildings for churches, doctor's offices, depots, city governments, and recreational halls. In Springhill the governmental meetings and church meetings were held in the same building.

In 1900 Buchanan built the Black Mason Hall located on what is now Main Street. It was used to house the first African-American school. He brought into town Mr. John Roseburough to help J. F. Giles with the bookkeeping at the mill. Mr. Roseburough became the first principal of the school in 1901. Gertrude Smith was his assistant.[7]

Early African-American Education in Webster Parish. Mrs. Mae Dee Moore, a retired Webster Parish teacher, recalls the early days of African-American education in Webster Parish. "Early schools were operated in different communities by the action of people in the community. They built log cabins or boxed in school buildings. Classes were held in many churches. These schools were opened two or three months during the winter each year. Each community had its own African-American school board. Each community hired and fired its own teachers."

Two funds assisted the African-American schools in Webster Parish—the Jeanes Supervision Fund and the Julius Rosenwald Fund. The Jeanes Supervision Fund was confined to the education of rural African-American education in the South. The Julius Rosenwald Fund was used for the purpose of improving buildings of African-American schools in the South. Mr. Rosenwald gave $4,273,927 for this purpose.

Brown High School. The first African-American high school was constructed in 1948 and named Springhill Colored High School. There were eight faculty members including the principal, Mr. Charles Herbert Brown. The school had grades 1—11 based on a term lasting eight months. Mr. Brown served as principal of the school from 1948 until his death March 3, 1951. He had numerous offers from schools in the East, but he chose to remain in Springhill and give back to the children under his care what others had done to further his education.

After his death the name of the school was changed to Charles H. Brown High School. This occurred during the 1951–52 school year.

Mr. Candalie La Mourne Capers was the next principal and served 1952–53. Mr. John Coleman became principal in 1953 and served until 1970. After the integration of Brown High School, Mr. Coleman went to the Webster Parish School Board central office and served as elementary advisor. He retired in 1986.

Integration

The integration of schools spread over the South after it was initiated in Little Rock in 1957. The Webster Parish School System had been challenged to integrate by a Court Order in 1965. Louis Padget, Webster Parish District Attorney, met with the school board to discuss a suit that had been filed to integrate facilities in the parish. Integration was delayed.

In 1970 Judge Skaggs ordered a plan for the integration of schools in the parish. Four plans were eventually used to comply with the court order. Minden used a "zoning plan." Most schools fell under a "consolidated plan." Doyline used a "pairing plan." Springhill used a combination of the "zoning plan" and the "consolidated plan." Brown, previously an African-American high school, was changed into an elementary and junior high school housing grades one through eight. Students were zoned into Brown, Browning Elementary, and Springhill Junior High if they were in grades one through eight. Springhill High School was converted into a high school for African-American and Anglo-American students.

Teachers were transferred in order to carry out the 62%–38% ratio of White and African-American teachers. The court order required 38% of the teachers be African-American. This level was to be maintained.[8]

Responses to integration. In a response to the question, "Has the school board solved the problem of integration?", Richard Noles, Superintendent, responded, "I believe we have turned the corner in all problems related to it. We're moving ahead. It's still up to the students to take the opportunity."[9]

One public official said, "I don't think either side wanted it, but we have learned to live together since both sides inhabit the same earth."

When integration was a definite policy for Webster Parish schools, an elaborate awards banquet was held at Brown High School. Teachers were exchanged. Brown was changed from a high school to an elementary school and junior high school for grades one through eight.

When the day arrived for students to leave Brown High School and transfer to Springhill High School, some students wept openly. The principal, Mr. Coleman, escorted the student body to the new school, Springhill High, where Donald Curry, the president of the Brown High School Student Council, made an acceptance speech on February 5, 1970. He said to his fellow students, the faculty, and the administration:

To the faculty and students of Springhill High School:

> We appreciate the manner in which you are receiving us into your midst. We realize these are difficult times for you just as they are for us. When we made the

Closeup of the Springhill High School mural.

Ivy-covered façade of Springhill High School in 1948.

Charles H. Brown School in 1948.

ALMA MATER
~Hodges
Memories of thy ivy wall live in us for ever
For we're bound by loyalty nothing else can sever

For the friendships we have known in these halls we call our own
We will love thee till we die Alma Mater Springhill High

Gymnasium/Auditorium in 1948.

Primary School Building for grades 1–3 in 1948.

New Springhill High School with additions of auditorium at right and classroom wing at left.

Elementary School Building for grades 4–7 in 1948.

School band marches past the Springhill Municipal Ice Plant and Cullen/Springhill bus stop.

Howell Elementary School, named in honor of Miss Georgia Howell.

Browning Elementary School, named in honor of M. T. Browning.

M. T. Browning.

Miss Georgia Howell.

Charles H. Brown Middle School, 2001.

Northwest Louisiana Technical College at corner of 1st St. N.E. and Haynesville Highway.

Chapter 8

Great Pines–Government Growth

City Government

Lawrason Act. The law was enacted in 1898 and named after Judge Samuel McCutchen Lawrason, a West Feliciana Parish lawyer born in 1852 in New Orleans. He received his law degree from the University of Louisiana in 1874. When he moved to West Feliciana Parish, he was elected parish judge, school board member, LSU Board of Supervisors, state senator and Vice-President of the Louisiana Constitutional Convention of 1898.

The act which bears his name continues to set the framework and guide the work of more than 75% of the incorporated municipalities in Louisiana. The intent of the law is to provide a uniform type of government for all municipalities of Louisiana.[1]

Springhill city government reflects this law. The law requires municipalities to be governed by the Mayor-Board of Aldermen form of government. The officers shall be a mayor, alderman, chief of police, tax collector and clerk. The mayor and police chief shall be elected at large. The number of aldermen in the city shall be not less than five nor more than nine. The number of aldermen in a town shall be five.[2] Springhill is defined as town and has five aldermen. Those serving at the time of the writing of this book are Ed Bankhead, Carroll Breaux, Richard Baker, Robert Hilburn, and Denny McMullan.

Mayor—City Council. The mayor's powers, duties, and responsibilities are listed in the law under nine functions, including the supervision and administration of all municipal departments, offices, and agencies.[3] John D. Herrington is serving his third term as mayor.

The aldermen have the responsibility of representing the people by the enactment of laws in the form of ordinances, appropriations of funds, incurring of debt, issuance of bonds, and development of budget.

The clerk keeps books labeled "Municipal Minutes of the Town of Springhill," and "Municipal Dockets of the Town of Springhill." The clerk is the custodian of the municipal seal. Other records of ordinances and land sales are kept by the clerk. Jimmie N. Murph is serving in this capacity. Tax assessments are made by the clerk or tax collector by copying from the assessment rolls that portion which embraces property or persons within the corporate limits.

Chief of Police. He has general responsibilities for law enforcement in the municipality, and is charged with the enforcement of all ordinances within the municipality and all applicable state laws. He shall perform all other duties required by ordinances. He shall make recommendations to the mayor and board of aldermen for the appointment of police personnel, promotion of officers, and disciplinary action. Sherrel Smith is Chief of Police.

City Court. The Lawrason Act requires a mayor's court in the municipality with jurisdiction over all violations of municipal ordinances. The act provides for an attorney appointed by the city council upon the request of the mayor. The city attorney tries breaches of the ordinances and imposes fines and/or imprisonment. John Slattery, Jr. is the City Attorney.

Ordinances

Code of Ordinances. An ordinance is defined as a statute or regulation, especially one enacted by a city government.[4] Springhill has a *Code of Ordinances,* a book listing all the ordinances passed by Mayors and Aldermen, past and present. It is the guideline for city government.

The *Code* embraces only the ordinances of a general and permanent nature. All ordinances designated special ordinances of the city are expressly saved and not affected by the adoption of this code.

Specific statutes. The *Code of Ordinances* is composed of 122 chapters, preliminary materials, and appendix. Many chapters are further sub-divided into Articles and Sections. Certain chapters are reserved for new ordinances and amended ordinances.[5]

The other chapters are designated for statutes or regulations in the following categories: General Provisions, Administration, Alcoholic Beverages, Amusements and Entertainment; Animals, Aviation, Buildings and Building Regulations, Businesses, and Cable Communications; Cemeteries, Civil Emergencies, Elections, Fire Prevention and Protection; Floods, Franchises, Health and Sanitation, Human Relations, and Law Enforcement; Natural Resources, Offenses and Miscellaneous Provisions, Parks and Recreation, Personnel, and Planning; Solid Waste, Streets Sidewalks and Other Public Places, Taxation, Traffic and Vehicles, and Utilities; Vegetation, Vehicles for Hire, and Zoning. Springhill follows these guidelines for city government.

City Service Departments

Fire Department. The history of the Fire Department is centered around three concepts: separation, ice house, and growth. In 1991 the department was separated from the city government. A Board of Commissioners was established to elect the Fire Chief and to be responsible for the policies of the department.

The present chief is David Camp, the only paid employee, and the youngest chief in its history. The office is located in the historic ice house. Vehicle buildings are constructed around it. The department has 13 volunteer firemen who have in-house training, 4 fire trucks, 2 brush trucks, and 1 rescue truck. In the year 2000 the department had 160 calls to fires, 42 of them houses and 51 grass fires. Land has been purchased and plans have been made to build a new fire station on 3 acres at the end of Machen Drive on Percy Barnes Road.[6]

Water Department. The necessary liquid required for life is water. The Springhill City Water Department takes excellent care of the water system for the citizens. Eight people work full time to keep the water flowing and the sewer draining. Department personnel work on water meters, sewer lifts, leaks, sewer pumps, and water wells. There are five water wells in the city that pump water for citizens' use and consumption.

The Water Department is currently led by Solomon Porter since 1998. Others who have served the city as Water Superintendent are John Holland, Roy Tyler, Ray Pixley, Therral Farrell, and Steve Robinson.[7]

Police Department. The Lawrason Act of 1898, which sets the framework and guides the work of incorporated municipalities of Louisiana, states that the Police Chief shall be elected at large. He has responsibility for law enforcement in the municipality and all applicable state laws. He makes recommendations to the Mayor and the Board of Aldermen for the appointment of police personnel.

On June 1, 1999, Sherrel Smith began his term as Chief of Police for Springhill, Louisiana. He is one of seventeen personnel in the department. Since Chief Smith has been in office, the department has obtained thousands of dollars from grants. These have been used to improve training programs, upgrade drug enforcement efforts, purchase several items of equipment for police cars, and conduct surveillance of drug pushers. The police department is constructing a new 3000 square foot facility that will house 14 inmates and provide offices for the police personnel. Plans are to begin the project in April 2001.

There have been four Chiefs of Police in the Springhill Police Department: Rubin O'Neal, Tommy Jarrett, Jerry Stephens, and Sherrel Smith.[8]

Street Department. L. D. Nail, Superintendent of the Street Department, and his personnel have the job of keeping streets repaired, resurfaced, and free of debris. Since 1991 there have been three programs to improve streets in Springhill. In November 1991 citizens approved $2,200,000 in General Obligation Bonds Series 1992 for street improvement. Sixty streets were re-surfaced in three phases from 1993 to 1997. The Police Jury resurfaced four streets in Meadow Creek as a result of citizens' petitions.

In 1995 a grant was received to build a new bridge over Little Crooked Creek and a new bridge over 3rd Street at a cost of $1,000,000. In January 2000 a grant for $20,000 was received to resurface the parking lot at the Community Activities Center.

In the early days of Springhill (1937) Main Street was filled with potholes and mud. It was completely rebuilt and paved in 1948. Through the years administrations have been challenged by the pot-holed, rough, mud filled streets. They have met the challenge and have provided good streets.

Major Improvements

Springhill mayors and city councils have been responsible and visionary about improving services to meet the needs of citizens. Water, fire, police, and street services are the obvious ones, but additional improvements have been made by the city government that are consistent with the growth of the town.

A pioneering step in the realm of ordinances in the State of Louisiana may serve as a model for other municipalities. One Springhill City Council adopted an ordinance that requires a developer to construct water and sewer lines at his own expense. The city refunds money used for these purposes on a monthly basis after the house has been connected to the meter.

Streets. Street pavement and improvement is an ongoing task for each administration. The city acquired considerable equipment for street construction and repair. A program has been initiated whereby property owners pay for new streets, but the city personnel do the work. When the paper mill construction was begun in 1937, streets were in poor condition. Through the efforts of many mayors and city councils they have been improved.

Mail delivery. Home mail delivery was accomplished in 1955 during the administration of Charles McConnell. The City Council undertook the job of marking all streets, numbering all houses and businesses, and renaming some streets in order to get home mail delivery.

Accounting. A system of accounting, auditing, and taxation was installed many years ago when Ed Shultz was mayor. This decision has saved the city money and provided a systematic approach to handle all city funds. One administration sponsored a move to re-finance old city bonds. This change saved the city thousands of dollars in interest.[9]

Recreation program. Springhill is especially proud of its recreation program. The City Recreation Council, with the support of International Paper Company, developed a program for all ages. A recreation building, swimming pool, and playground serve as a center for teenage actvities. Softball and baseball diamonds are well-lighted and maintained throughout the city.

City park. The city park was improved in 1944 with the addition of a children's wading pool. Other amenities were added in the following years such as children's playground equipment, picnic tables, and a new gazebo.

Ice plant. In 1945 the city began rebuilding the municipal ice plant. Electric refrigerators were not found in many homes during earlier days. It was a necessity for the "ice boxes," forerunners of the modern electric refrigerators, to have a supply of ice immediately available. The ice plant was the answer. Today, the office of the Fire Chief is located in the old ice plant building on Main Street.

Fire fighting equipment. Between 1946 and 1950 water and sewer lines were extended and enlarged. This provided a better source of water for fire fighting because water pressure was increased. A fire station was completed at a cost of $28,000 and a new fire truck was purchased for $5,700.

Frank Anthony Park. In the 1990s Mayor Herrington and the City Council developed ideas for Frank Anthony Park and the Community Activities Center. There are thirty RV camping pads with full hook-ups. Walking trails are in place in the area, as well as picnic tables and a barbecue pavilion. Joe Curtis is chairman of the committee that is working on the project.

Main Street Program. The Main Street Program promotes beautification of Main Street buildings, economic development, and special events. Jan Willis is director of this effective program.

Airport. A new airstrip has been paved which allows small jet aircraft to land at the Springhill Airport. This will provide an opportunity for corporate jets to bring executives to the area to visit their manufacturing plants or to visit sites for possible new businesses. Local pilots enjoy use of the facilities.

Lumberjack Festival. Through the co-operation of the Mayor, City Council, and Chamber of Commerce the Lumberjack Festival is promoted each October. The festival draws thousands of visitors to Springhill each year. The Festival was begun 1983 during the first term of Mayor Johnny Herrington.

Bus service. In the mid and late 1950s a bus service between Springhill, Minden, and Shreveport was planned and promoted by Mayor McConnell and the City Council. It was eventually discontinued because of lack of use. However, it serves as another example of a service provided by the Mayor and City Council to meet the needs of citizens.[10]

Recent City Projects

1999 Mayor's Report. Mayor Herrington presented a progress report of recent projects either planned, funded, or in progress in 1999.[11] These include Little Crooked Creek Flood Control; walking trail at Anthony Park ($40,000); pavilion and picnic tables, soccer field, and landscaping at Anthony Park; National Guard work in the park; airport erosion control, renovation of the old A & P store for development into the Community Activities Center, with a building grant ($400,000); sidewalk improvement from Church Street to City Park; trash receptacle grant ($20,000); bricks for sidewalk in the historical area; airport runway extension ($150,000); new police station; Centennial Committee 2002; Clock Monument for Main Street; Leadership Grant ($5,000); and Main Street Program ($8,000).

These projects are in the historic tradition of other mayors and aldermen who demonstrated in their time a progressive spirit and a constructive vision.

2000 Report. Mayor Herrington elaborated on other basic improvements in his 2000 Report. In November 1991 citizens voted for propositions that would fund street and sewer improvements.

A sum of $2,200,000 in General Obligation Bond Series 1992 was approved for street improvements that will be completed in three phases. Another bond issue was approved for a 1% sales tax to fund $6,300,000 for the construction of sewer improvements. It will be enough to finance 16 new lift stations, new sewer lines in the Pine Hill Subdivision, new sewer lines on the east and west sides of Springhill, and the rehabilitation of the wastewater treatment plant. Sixty streets have been resurfaced.

In 1995 the city received grants from the state and federal government to clean out Little Crooked Creek and build a new bridge over Third Street N.E. In 1994 the Louisiana Community Development Grant for $507,370 was awarded to the city to renovate houses on Third and Fourth Streets S.W.

Police Jury[12]

Louisiana is unique in the nation in that it has parishes which are governed in most cases by police juries. At one time Louisiana had counties. In 1804 the Legislative Council divided the state into 12 counties which were found to be too large for satisfactory administration. In 1907 the state was divided into 19 parishes that corresponded to the 21 ecclesiastical parishes established in 1762.

Parish government was adjusted between 1807 and 1830 to require member juries to serve with the parish judge to administer the affairs of the parish. The office of sheriff and the wards were defined. In 1811 the legislature created an elected police assembly officially designated as a "police jury."

The 1974 state constitution granted broad home rule authority to parishes and municipalities and reversed the traditional concept of local government as a "creation of state" possessing only delegated authority.

The police jury system vests both legislative and administrative functions in the same body. The legislative functions include enacting ordinances, establishing programs, and setting policy. The police jury performs administrative functions of budget preparation, hiring and firing personnel, spending funds, and directing activities under its supervision.

Police Juries exercise over 50 functions, including, but not limited to, road and bridge construction and maintenance, drainage, sewage, solid waste disposal, fire protection, recreation and parks, parish prison construction and maintenance, and health units and hospitals.

Webster Parish has 12 districts, thus 12 members. Tylon Blanton, Jimmy Thomas, and Daniel G. Thomas represent Springhill on the Police Jury.

Wall of Mayors inside City Hall.

Sign at southern entrance to Springhill.

Above: *New street lights graced Main Street in 1948.* Inset: *Citizens await the parade celebrating the new paving project.*

Above: *S.H.S. Band parading south down the new Main.*

Left: *Governor Earl K. Long addresses citizens at the celebration for a newly paved and lighted Main Street. Mayor Ed Shultz seated at right.*

City Hall flanked by fire station and Masonic Lodge south of the city park in 1948.

Community House with Boy Scout Hut (left), and wading pool, in 1949, at site of current Civic Center.

Right: *Close-up of Christmas tree in the triangle at north end of Main.*

Below: *First municipal Christmas decorations about 1948, looking north on Main toward Christmas tree.*

Construction scene extending Main Street toward Cullen in 1948.

Springhill Ice Plant in 1945, open 24 hours a day.

Current fire station encompassing the old municipal ice house.

Fire station built across from City Park about 1948 (demolished in 2001).

Left: *Springhill Airport about 1944.*

Below: *Aerial view of airport southeast of town.*

Clory Haynes Allen, librarian, in the former parish library.

Current Webster Parish Library, east side of Main.

West side of Civic Center as seen through the pines of City Park.

East side entrance to City Hall.

Chapter 9

Great Pines–Medical Growth

History of Medicine

Primitive medicine. The history of medicine and surgery is the account of man's efforts to deal with disease and human illness, from the primitive attempts of preliterate man to the present complex array of specialities and treatments.

It seems probable that man, as soon as he reached the stage of reasoning, discovered which plants were good for food, which ones were poisonous, and which of them had some medicinal value.

Early history. About 3000 B.C. the Code of Hammurabi, law code of an early king of Babylonia, included laws relating to medical practice. Hebrew literature, particularly the Bible, contains little on medical practice, but an abundant amount of material on personal hygiene.

Hindus believed the body consisted of three universal divine forces: spirit, phlegm, and bile. Health depended on their normal balance. The great aim of Chinese medicine was to control in the body the proper proportions of yin and yang.

Greeks followed Hippocrates on the island of Cos. He practiced holistic medicine: an emphasis on body, mind, and soul. His greatest legacy was the Hippocratic oath adopted as a pattern by medical people throughout the ages and is used in graduation ceremonies at many schools of medicine today.

Greek and Roman. The Renaissance movement of the 14th, 15th, and 16th centuries revived an interest in Greek and Roman culture, an escape from tradition, an eagerness for discovery, and an exploration of new fields of thought and action. In medicine anatomy and physiology were the fields of inquiry that grew during this time.

Later history. William Harvey in 1616 laid the foundation for modern embryology. His discovery of the circulation of the blood was a medical landmark. Louis Pasteur proved in the 19th century that fermentation and disease were caused by living organisms known as bacteria.

The 20th century was one of great discovery and advances in all fields of medicine: chemotherapy, immunology, surgery, endocrinology, and nutrition.[1]

Early North Louisiana Medicine

Caddo medicine men. The first medical care in North Louisiana was given by the Caddo medicine men who let nature take its course in most cases.[2] According to one Indian theory, illness was caused by the spirits of animals. Other diseases were caused by human and animal ghosts. Doctors (medicine men) treated by using a combination of herbal remedies and singing or reciting magical formulas.

Magical religious power and the gift of healing which gave the medicine man authority came from visions. His visions showed him objects to put together as a charm that would protect him and bring him a good life. These charms were called "medicines."

Early drugs. Since doctors were scarce in rural areas, the early settlers often used natural medicines. They used flowers, leaves, fruit, roots, and seeds to alleviate and cure ailments. Sweat cloths, castor oil, and calomel were used; salt was used for ear aches; tobacco was chewed and applied to insect bites. Alcohol and aloe vera plants for burns were also common medicines used by settlers. Some people chewed willow bark for pain relief. Indians used the foxglove plant for hundreds of years before scientists discovered its heart-treatment effects. Garlic was a remedy for high blood pressure.

Early diseases. At the beginning of the 19th century Northwest Louisiana was considered quite healthy because of sparse population and isolation of the people, but as population density increased the number of diseases continued to grow. The two great epidemic infections of the 19th century were yellow fever and Asiatic cholera. Although they were feared by the settlers, neither one proved to be a serious threat to Northwest Louisiana until the mid century.[3]

Tuberculosis was a major killer disease in the 19th century. Malaria, enteric disorders, and respiratory infections were the major problems of North Louisiana.[4] The endemic diseases of the Western world such as diphtheria, measles, scarlet fever, and typhoid occasionally appeared, but the relative isolation of the communities in Northwest Louisiana saved them from the worst effects of these infections. Early Springhill residents did suffer from flu, pneumonia, whooping cough, small pox, and measles, however.

Early doctors. The first Anglo-American doctor to settle in Northwest Louisiana was John Sibley who lived on a plantation near Natchitoches in 1802–1803. In 1804, President Jefferson appointed him Indian Agent for the Territory of Orleans.

Early physicians were influenced by the British schools of Oxford and Cambridge whose graduates catered largely to the upper class. Few migrated to the United States which resulted in a constant shortage of trained medical men in America. As late as the American Revolution in 1776 only ten percent of the physicians in the colonies held medical degrees.[5] Most of the early doctors in Northwest Louisiana were educated at Tulane Medical School or else they served as apprentices to practicing doctors for one to twelve years to receive their training.[6]

Dr. Andrew Jackson McDonald. The first licensed and graduate doctor to serve Springhill was Andrew Jackson McDonald who graduated from the University of Arkansas Medical School.

While he was practicing medicine with Dr. Bond in Shongaloo, he was contacted by Mr. Harris, manager of the newly constructed Pine Woods Lumber Company. In

1898, Bond asked Dr. McDonald to come to Springhill several times each week to check on employees who were ill or had been injured.

In 1898 he agreed to come twice a week. He rode his horse from Shongaloo to Springhill until he moved his family permanently in 1900. When he arrived in Springhill, he opened an office to practice medicine and a general merchandise store.[7] He practiced in both Arkansas and Louisiana until 1945.

Dr. Joseph Robert Browning. Joseph Robert Browning was one of four children of George W. and Mary Jane Bross Browning. He was born March 1867 in Claiborne Parish, Louisiana. The Federal Census of 1900 placed him in Webster Parish.

He practiced medicine in Springhill from 1900 until his death November 21, 1927. On March 3, 1895, he married Della Lindsey with whom he had three children. Two became doctors and a daughter was an accomplished musician, music teacher, and church music director.

Dr. Rupert Butler. Dr. Rupert Butler attended Memphis Medical School in Tennessee for twelve years to train for the medical profession. He was graduated in 1905. He married Katherine Goodwin in Red Land, Louisiana, in 1907 and the newly married couple lived near the Salem Baptist Church.

Pine Woods Lumber Company sent Tom Blunt to Red Land to ask Dr. Butler to be the mill doctor. He accepted and moved his family by wagon to Springhill where he and Kate bought a lot for their home on "the hill." In later years the city named the street "Butler Street" in Dr. Butler's honor.[8]

Dr. Butler made house calls, sometimes spending the night at the bedside of a seriously ill patient. According to his daughter's observation Dr. Butler's primary purpose was to practice medicine and take care of his family. He would not refuse anyone a house call day or night. He was on call every day of the year. Because of his devotion to helping others by practicing medicine, he was loved, respected and admired. Local citizens paid him a tribute by naming the Webster Parish health clinic in his honor—Butler Memorial Health Center.

Dr. Charles Terrell McWilliams. Dr. McWilliams was a native of Arkansas who practiced medicine in Springhill. After he earned his pharmacy degree and his medical degree from the Kentucky School of Medicine and the University of Kentucky in 1910, he practiced medicine in Columbia County, Arkansas, until he moved to Springhill in 1937. In Columbia County he went to see his patients on horseback. In Springhill he practiced with Dr. Garrett at Swan Clinic. He was known for diagnostic work.

Butler Memorial Health Center

Plaque. The brass plaque on the front of the Health Center reads: "In memory of Dr. Rupert Butler, M.D. (1879—1948), who unselfishly served the community for 31 years –Webster Parish Police Jury 1949."

Concept and development. The original concept of a health center came from Mayor Ed Shultz and the Springhill City Council. They presented the idea to the Webster Parish Police Jury and encouraged them to complete the project. The Police Jury completed the building in 1949. Members of the Police Jury who developed the project were J. M. Pearce, President, J. L. Mann, K. O. Martin, E. W. Brown, C. A. Slack; W. E. Stewart, G. C. Garland, W. M. Love, and F. B. Treat. The Health Center is located at 218 1st Street N.E. The Health Center is known by two names: Butler Memorial Health Center and Webster Parish Health Unit.

Funding. Money for programs is funded by the federal and state governments. The building budget is funded by the Webster Parish Police Jury.

Programs. The center has numerous programs for everyone—some for the indigent and others for all persons who pay on a sliding scale basis.

These programs include family planning, maternity care, child health care, WIC vouchers for food, and STL; clinic, vital records, epidemiology, tuberculosis screening, and flu shots. Two office personnel, a full-time nurse, and two part-time doctors manage the health unit.

New building. A one-acre tract of land has been purchased by the Police Jury to construct a new building near the Frank Anthony Park on Church Street. Although the old building was remodeled and doubled in size in 1975, more space is needed. The project is being encouraged by Police Jury member Tylon Blanton.

Hospitals and Clinics

Swan Memorial Hospital. In 1938 when the paper mill began operations, Dr. Garnea moved to Springhill to become the doctor for the employees of the newly-constructed mill. One of Dr. Garnea's first projects was to build the first hospital in Springhill. He built it exactly like the one in Bastrop and named it for Mr. Swan, a close friend who had been killed in an automobile accident. Swan Memorial Hospital was constructed in 1938.

The first floor had space for an x-ray machine, scrub room for surgery, surgical room equipped with small boilers that were used to sterilize instruments and clothing, patient examining rooms, and a laboratory. The second floor consisted of hospital rooms for the ill.

Garrett's Clinic. In 1943 Dr. William Garrett purchased the Swan Memorial Hospital from Dr. Garnea and changed the name to Garrett's Clinic. Dr. Wilson Gray was a partner with Dr. William Garrett, but soon dissolved the partnership in order to build his own hospital.

Gray's Hospital and Clinic. When Dr. Garrett and Dr. Gray dissolved their medical partnership, Dr. Gray built his own hospital and clinic and named it Gray's Clinic. It was located at the present site of the Sunbridge Care Center-Fountain View on 1st Street N.E.[9] Dr. Gray continued to practice and kept his hospital open until the new hospital was completed in 1975. At that time he closed his hospital and admitted patients to the new hospital.

Sims Clinic. Dr. Howard Sims left a thriving practice in Bastrop, Louisiana, in order to practice in Springhill where a new International Paper Company mill began operations in 1938. Dr. Sims moved to Springhill in 1937. People who moved from Bastrop knew Dr. Sims, so his medical practice grew rapidly when these same people transferred to Springhill.

This physician had attended the University of Louisville Medical School in

Kentucky. His wife was a nurse and assisted him with his work. Mrs. Sims was an honor graduate of University of Louisville School of Nursing. Together they built the Sims Clinic on the South Arkansas highway, halfway between the town of Springhill and the new paper mill. Medical facilities were on the first floor and the family lived on the second floor.[10] Dr. Sims was known for his excellent work in obstetrics.

Two other physicians who practiced medicine during the early years of Springhill's history were Doctors Lewis and Wall. Dr. Lewis worked with Dr. Gray and Dr. Wall worked with Dr. Garrett.

Doctor's Clinic. Dr. E. A. Hand founded a clinic in the early 1950s. It was located in the second block of 1st street N.E. Dr. Marvin Soileau practiced with Dr. Hand in 1955 while he was an intern. He became a partner in 1957. Dr. Hand built a hospital and clinic in 1957 across the street from their original clinic. It was located adjacent to Northwest Louisiana Technical College. They named it Springhill General Hospital.

Other doctors began to join the Hand-Soileau staff in the old clinic: Dr. Sam Holladay 1959, Dr. Charles Payne 1969, Dr. Wayne Sessions 1976, Dr. David Law 1976, and Dr. Charles Hunsinger 1982; Dr. Gary Torrence 1987, Dr. Marlin Morris 1960, Dr. Archie Robinson 1959. Dr. David Law rejoined the clinic in 1993, and Dr. Steve Ditto 1997, Dr. McClain 1997, Dr. Mahamed Ahmed 1999; and Dr. Cecilier Chen 1999. At the time of writing (December 2000) Doctors Ahmed, Edwards, and Karma are no longer with the clinic. Dr. Soileau retired effective October 31, 1999.[11]

Springhill General Hospital (downtown) was sold to Extendicare, Inc., in 1969. Extendicare later became Humana Hospital. Extendicare bought the old hospital (downtown) with the proviso that they would build a new one within five years. Dr. Soileau and Dr. Holladay sold them the land for the new hospital. The doctors reserved two acres on which to build a new clinic that became known as Doctors Clinic. Extendicare built a new hospital in 1975 and closed the old hospital. The hospital, located at 2001 Doctors Drive, was named Springhill Medical Center.

The Doctors Clinic stated their role in Springhill medicine in their publication:

> For more than forty years Doctor's Clinic has provided medical direction and leadership for the community. Physicians at the Clinic serve as medical staff for the Springhill Medical Center while providing medical support to the entire community. The Clinic provides more than $20,000 worth of free physicals to the area high school athletes each year. Clinic doctors can be observed on the sidelines for all Springhill High School football games. The physicians serve as medical director for family planning at the Public Health Center, for area nursing homes, and for rural health centers. They also provide industrial medical care to local manufacturing and commercial ventures.[12]

Springhill General Hospital

Humana Hospital. Springhill General Hospital closed and moved to the newly-constructed Humana Hospital in 1975. It was operated by Humana through 1987-

1991. Dr. Holiday, Dr. Soileau, Dr. Payne, and Dr. Morris were the physicians practicing in the hospital. It was then sold to Columbia Medical Corporation.

Columbia Hospital. Columbia purchased the Springhill General Hospital from Humana in the early 1990s. By the mid 1990s financial trouble was so intense within the Columbia system that it became a national issue. When the business difficulties occurred, Columbia sold over 100 of its 400 hospitals across the nation. Springhill remained within the Columbia system.

The primary purpose was to make money for the corporation. At the same time the Doctors Clinic physicians operated on a patient-care medical philosophy. Tension grew when Columbia management began sending $82,000 per month off the top of profits to the Nashville main office thus effectively raising the stock prices from $7 to $35. The Clinic/Hospital relations were generally very good and fees were reasonable until Columbia began to go into competition with the Clinic in the mid 1990s. They did so by opening Rural Health Centers in Taylor, Cullen and Cotton Valley. Doctors were recruited in specialty areas of OBGyn, orthopedics, ophthamology and neurology to come from Shreveport weekly to serve the Springhill community.

Life Point. Columbia sold the Springhill Community Hospital to Life Point Corporation.

Springhill Medical Services. Greg Simmons, Jimmy Robertson, and Dennis Robertson organized a local health-care company entitled Springhill Medical Services, Inc. Through this company $8 million was borrowed from Bank One of Shreveport to purchase the 63-bed Springhill Medical Center from Life Point. The loan was secured by the United States Department of Agriculture guaranteeing 90% of the loan. The other 10% came from investors as earnest money. Ten investors will make up the board for the non-profit medical center.[13]

The board members for the new local hospital corporation are Dr. Wayne McMahen, Thomas Garland, Dr. Raymond Robertson, Dr. Don Teague, and Dr. Marvin Soileau; Richard Threet, Dr. Sam Holladay, Jr., Jerry Dale Thompson, Dennis Robertson, and Dr. Gary Torrence, John Dennis, Johnny Herrington, Bob Colvin, James Robertson, and Bob Bush. Greg Simmons is the new chief executive officer of Springhill Medical Center.

The corporation that bought the medical center was formed July 17, 2000, and named Springhill Medical Services, Inc. It is a 501-C-3 non-profit corporation. Documents transferring the Life Point Hospital to Springhill Medical Services, Inc. occurred November 15–17, 2000.

The Springhill Medical Center will operate under a patient-care philosophy that matches the Doctors Clinic philosophy of medical practice. In January 2001 a total evaluation was made of all costs within the medical center with the goal of "fairness for everyone." On December 18, 2000, two doctors from Doctors Clinic and the new CEO of Springhill Medical Center visited medical schools and hospitals to recruit new physicians.[14]

"Only after being approached by Springhill Medical Services did Life Point Hospitals, Inc. consider selling Springhill Medical Center," said Jim Fleetwood, Jr., chairman and CEO of Life Point Hospitals, Inc. "We took a close look at the needs of the Springhill community and performed the usual due diligence activities that accompany this type of transaction. After a thorough review, we determined that selling the hospital to Springhill Medical Services, Inc. was the best decision for the community of Springhill."[15] Good relationships and happy experiences are anticipated between doctors, nurses, Board of Directors, and C.E.O. because of the patient care philosophy and the non-profit status of the hospital.

Sunbridge Care Center—Fountain View

Professional health care service. Sunbridge Care Center—Fountain View is a 153-bed facility overlooking the city park located within two miles of the local hospital and Doctors Clinic. The care center provides quality rehabilitation therapy and professional health care services for short term and long term residents. Recreational and social opportunities include beauty services, cooking groups, games, field trips, and respite care in a home-like environment.

Medical services are offered to patients recovering from hip replacement, knee surgery, or other orthopedic conditions, traumatic brain injuries, strokes; open heart surgery, and surgical wounds.

Rehabilitation therapy services help individuals achieve their highest level of independence. Experienced therapists follow a treatment program prescribed by a physician for physical therapy, speech therapy, occupational therapy, and respiratory services.[16]

Ownership. In 1964 Dr. Gray went into partnership with Mr. L. M. Cadenhead to establish Fountain View Care Center located within the former Gray Hospital. However, he continued to practice medicine by keeping his hospital open until the new hospital was completed in 1975. At that time he closed his hospital and admitted patients to the new hospital. He died within a year of the new hospital opening. Healthcare Capital purchased the facility in 1995 from the Cadenheads. Sunbridge Healthcare Corporation became the owners in 1997. On December 20, 2000, Fountain View became known as Fountain View Care and Rehabilitation. HQM, Home Quality Management of Palm Beach Gardens, Florida, became the new owners.

Susan Kottenbrook is the administrator. Dr. Marvin Soileau serves as Medical Director with a staff of 150 people. Other doctors requested by the residents provide medical care as needed.

Dentists

Dr. Matthew Lane. Dr. Lane is a 1984 graduate of Louisiana State University Dental School. His first practice was in Rayville in 1984–1985 with his brother. He came to Springhill in 1985 to practice for a brief time with Dr. Gary Chumley. In August 1985 Dr. Lane began an independent practice in a new facility. Dr. Matthew Lane now has five employees.[17] He is a member of the Louisiana Dental Association, American Dental Association, and Ark La Tex Academy of Dentistry. His dental emphasis is upon general practice and cosmetics.

When the author asked him why he came to Springhill, he offered a spiritual answer, "This is where God had me come. His hand was on it." Dr. Lane is an active member of Central Baptist Church where he participates in the choir and serves as a deacon.

Dr. Charles Tanner. Dr. Tanner is a 1953 graduate of Baylor Dental School in Dallas. He has been practicing dentistry in Springhill since his graduation in 1953. His office was located in the Springhill Bank Building for five years. In 1958 he moved to his present location on 1st Street.

Dr. Tanner is a native of El Dorado, Arkansas. At the age of 12 he moved with his family to Homer, Louisiana, where he graduated from high school. He is a Sunday School teacher and an active member of Central Baptist Church. He is a member of the American Dental Association, the Louisiana Dental Association, and the Northwest Louisiana Dental Association.[18]

Dr. Jason Robertson. Dr. Robertson graduated in 1993 from the University of Tennessee Dental School. He practiced in Memphis one year before he opened his office in Springhill in December 1995.

Dr. J.E. Rutledge. Dr. J.E. Rutledge received his dental degree from Loyola University in 1943. He served in the United States Dental Corps four years 1943–1947. At that time he returned to Springhill where he practiced dentistry 41 years 1947–1988.

Other dentists who practiced in Springhill in the early years were Dr. Pacey, Dr. Young, Dr. Edward Lee Harris, and Dr. Hodgkiss.

Pharmacies and Pharmacists

Drug stores. Throughout her history Springhill has been blessed with good drug stores that provided medicines for the community. Some have been established, served well for many years, and then closed. Other stores are new in town. Their names are familiar to some and unfamiliar to others: Bond Drug Store, Tennyson's Drug Store, H & B Drug, Roy's Pharmacy, Fred's Pharmacy, Mall Pharmacy, Robertson's Family Pharmacy, and Robinson's Drug Store.

Tennyson's Drug Store. Murray Tennyson established his drug store in 1937 and located it in the Springhill Bank Building at the corner of Ensey and Giles Streets where the drug store remained until 1940. It was closed in 1996 when Olin Mills, the pharmacist, retired. The drug store was moved from the Ensey and Giles location to a new location in 1940 at the corner of Main and Church Streets. Merchandise was hauled from the old location to the new location in a wheel barrow.

In addition to excellent service Tennyson's Drug Store became known for its cherry cokes for the young people and coffee for the adults. Each week there would be an advertisement in the *Springhill Press and News Journal* announcing that Tennyson's Drug Store had "the best coffee in town." Each afternoon when the school day was completed one could see several bicycles parked in front of the store while young people enjoyed their "cherry cokes." The fountain was a busy place in the 1940s and 1950s.

The store delivered fountain foods and drug store medicines to customers on bicycles. One lady ordered a 10 cent coke daily that was delivered to her home. Obviously, the profit margin was small on that item. Curb service was also provided for customers.[19]

Olin Mills was the pharmacist at Tennyson's Drug Store 1950–1996. He began working for Murray Tennyson when he completed his studies in pharmacology in 1950 at the University of Mississippi. After Mr. Tennyson died in 1957, Olin Mills bought the store from the Tennyson daughters. Except for two years in the army 1950–1952, Mr. Mills worked at Tennyson Drug Store until his retirement in 1996. He and his wife Dot were the parents of two sons. He has been an active citizen, a member of the Lion's Club, and a member of the Springhill United Methodist Church.[20]

H & B Drug. After finishing high school in 1944 A. C. was hired by L & A Railroad. That afternoon Mr. M. C. "Doc" Burnham told A. C. he would loan him money to attend pharmacy school if he would go to work for him. Higginbotham enrolled in Loyola in October 1944. His school was interrupted by service in the Army. He returned to Loyola and graduated there in May 1949.

Dr. A. C. Higginbotham worked as the pharmacist at Burnham's Drug until January 1950. At that time he transferred to Allen Pharmacy in Henderson, Texas. Then, in 1950 Jesse Boucher asked him to join in at Boucher Drug. In September 1950 the Higginbothams purchased one-half interest in the pharmacy which became H & B Drug. In 1962 they purchased the other half interest in the store.

It was a typical drug store with drugs, soda fountain, sandwiches, and lunch. In 1969 they decided to close the soda fountain and handle gifts and drugs. Dr. Higginbotham said it has been interesting to say through the years that they had the first wall to wall carpet in Springhill. In 1982 he returned to Loyola to receive his Doctor of Pharmacy degree. The store continued to be successful until his retirement in 1989.

Fred's Pharmacy. Suzanne Souter Smith is the pharmacist at Fred's Pharmacy. She is a 1993 graduate of the University of Louisiana at Monroe School of Pharmacy. She has been the pharmacist at Fred's since 1994.

Mr. Howard Moody is a 1963 graduate of University of Louisiana at Monroe School of Pharmacy. For 21 years he was pharmacist at Gibsons. He then served for 7 years in his private pharmacy located in the present day tobacco shop. In 1993 he sold to Fred's Pharmacy where he works part time.

Mall Pharmacy. James O'Glee is a graduate of Northeast State University School of Pharmacy. He purchased the pharmacy from Yates Calhoun on January 7, 1985. This pharmacy has been owned previously by Tom Barnard and Roy Folse. It was located at that time on the east side of Main Street north of present day Avalon Technology/Springhill dot Net. James O'Glee purchased the inventory from H & B Drug when that store was sold. He has expanded the Mall Pharmacy since he purchased it in 1985.

Robertson Family Pharmacy. Dr. Raymond Robertson opened his store, Robertson Family Pharmacy, in Springhill in 1978 and has practiced pharmacy since that time. Dr. Rhonda Harvey has practiced pharmacy since 1990 in the Robertson Family Pharmacy.

Optometrist

Dr. William Odom. Dr. William Odom earned his Doctor of Optometry degree from Southern College in Memphis, Tennessee in 1950. He has practiced continuously in Springhill since 1950, first in his office on 1st Street next to H & B Drug, then at his present office on Church Street since 1978.[21]

Chiropractors

Dr. Wayne Smith practices two days a week in Springhill. The other days he practices in Ruston. Dr. Heintze retired in Springhill, but now has a part-time practice in town.

Veterinarians

Dr. Royce McMahen practiced veterinary medicine until he had health problems. He then ran for Sheriff of Webster Parish. When he assumed responsibilities for the Sheriff's office, his son, Wayne McMahen, continued the family tradition of veterinary practice in Springhill.

Summary

Springhill has become a regional medical center for small communities in the surrounding region. Medical facilities and doctors are excellent. The new hometown corporation reflects a spirit of enthusiasm and positive feelings about the future of Doctors Clinic and the hospital. The Hippocratic oath is strong in Springhill.

Top right: *Dr. Rupert Butler.*

Bottom: *State Senator Drayton Boucher dedicates Butler Memorial Health Clinic. Mrs. Kate Butler seated rear center.*

A. C. Higginbotham in the years when all pharmacists compounded medicines. Note the hundreds of apothecary jars.

Swan Clinic, one block east of Main on Bice St. in 1948.

H&B Drug, Butler Memorial, and Gray Clinic.

Gray Clinic in 1948, facing north across from City Park.

Butler Memorial Health Center of Webster Parish.

Springhill General Hospital one block east of Main, now closed.

West side of Fountain View Care Center.

Springhill Medical Center east of Doctor's Clinic, 2001.

Doctors' Clinic on 11th St. N. E.

Current home of North Webster Ambulance Service on Church Street.

Former North Webster Ambulance Service, now demolished, site of forthcoming Butler Memorial Health Center on Church Street.

Chapter 10

Great Pines–Organizational Growth

Organizations in a society are like mosaics in art. By themselves they look like individual pieces of a puzzle. But together they blend into a colorful pallet that produces a beautiful picture. In Springhill many organizations function by themselves in their own world. However, when one views them together, they look like a total photograph of the community. These organizations have a purpose. They have been founded to accomplish some task for the betterment of the community, and each will be described.

Organizational Theory

"Organizations, whether business enterprises, large government agencies, labor unions, large hospitals or large universities, are, after all, brand new. A century ago almost no one had even much contact with such organizations beyond an occasional trip to the local post office to mail a letter," wrote the founding father of the science of management, Peter Drucker.[1] Today people organize to accomplish some goal or to have fellowship. Organizations are a part of the landscape of American society and of other highly developed countries. They are everywhere and serve various purposes.

The following organizations are active in Springhill: Ark La Partners in Genealogy, Beautification Committee, Civic Club, and Frank Anthony Park Committee; Habitat for Humanity, Lions Club, Lumberjack Festival Association, Masonic Lodge, and Quarterback Club; Riding Club, Rodeo Association, Rotary Club, Senior Friends, and Springhill Art League; Springhill Country Club, Springhill Senior Center, and Woodchoppers.

Springhill Organizations

Ark-La Partners in Genealogy. Cindy Hall called a group of persons interested in genealogy to an organizational meeting on September 28, 2000. The purpose of the meeting was to establish a local genealogical club to help beginners in the field initiate their family history projects. Charter members of the organizational meeting were Cindy Hall, Dollie Kauffman, Rose Bailey, Dana Cook, and Sam Martin; Ann Bonner, Gloria Jacobs, Patsy Roberts, Helen Caraway, and Betty Holloway; Carole Castleberry, Louise Pharr, Stella Godley, Wanda Mason and Ruby Wise.

The Ark-La Partners in Genealogy elected the following officers: Rose Bailey, President; Wanda Mason, Vice-President; Gloria Jacobs, Secretary; Cindy Hall, Treasurer; and Carole Castleberry, Historian.

The organization meets monthly on the 4th Tuesday. The purpose of the group is threefold: to hear speakers on the subject of genealogy, to gather information about beginning a family history project, and to learn about things they should not do in a genealogy project.[2]

Beautification Committee. "A thing of beauty is a joy forever," wrote Keats. This literary idea encouraged Marie Gillespie and Don Murph to begin a decoration program for Springhill in 1996. A Beautification Committee was organized to decorate the city during the Christmas season.

Jean Soileau, J. B. Edens, Margaret Edens, Sammy Stanford, Harlon Driskill and Nancy Driskill also contributed their time and abilities to this endeavor. Through the efforts of the members of the Beautification Committee the city has been decorated during the Christmas season. Christmas lights and decorations have been placed on the Civic Center, City Hall, utility poles along the highways, the Gazebo, and in the City Park.

The Beautification Committee has enlarged its vision and expanded its work to include decorations at the flag pole at Frank Anthony Park and trees and flower pots on Main Street. Additional Christmas decorations have been added each year.

J. B. Edens is president of the organization (2001). This committee of talented citizens has added beauty to Springhill through their decorations. Their efforts have improved the quality of life in the city nestled among the pine trees of Northwest Louisiana.[3]

Civic Club. The Civic Club of Springhill was founded in 1946. It is a service club whose members are women who work to improve various areas of life in Springhill. They have contributed financially to some major cause each year such as the schools, the library, and the Habitat for Humanity.

The Springhill Civic Club is one of three clubs that has remained a member of the Federated Clubs of the 4th District. It has been honored three times as the Outstanding Civic Club in Louisiana.

The club once sponsored the Charity Ball, a popular event in Springhill. Since 1983 they no longer produce a Charity Ball but the club recognizes the Outstanding Club Woman of the Year. The Civic Club meets once a month to hear speakers and have fellowship. Bobbie Peters is the current president.[4]

Dinner Theatre. A drama theatre for Springhill is in the "vision/dream" stage of planning at the present time. A group of people is interested in purchasing a vacant bulding on Main Steet that can be remodeled and used for theatrical productions.

A dinner theatre has been discussed as a possible format for dramatic productions. Both children and adult dramatic plays will be produced. Interested citizens believe this approach to drama would be a worthy addition to the cultural and recreational climate in Springhill.

Frank Anthony Park Committee. The Frank Anthony Park Committee was selected by Mayor Herrington in 1996 to oversee the development of the recreational vehicle park named in honor of Frank Anthony, C.E.O. of Anthony Wood Products, Inc. Mr. Anthony's father began the sawmill operations, but Mr. Frank Anthony expanded the business into numerous sawmills.

The Frank Anthony Park is located on the site where the sawmills of Bodcaw Lumber Company, Pine Woods Lumber Company, Frost Lumber Company, and Anthony Wood Products, Inc. stood since 1896. The city purchased that 25 acres for $25,000 to develop the park with RV hookups, utilities, and a pavilion that can be used by overnight guests. The old A & P building next to the park is being renovated with a new kitchen, bathrooms, and meeting space that can be rented for group activities.[5] Joe Curtis is chairman of the Frank Anthony Park Committee. He and other members are developing the park on a long range basis primarily through grants.

Habitat for Humanity.[6] The Habitat for Humanity is an organization affiliated with the national group called Habitat for Humanity. Its purpose is to build houses for families who cannot afford to construct them by themselves. The owners must be approved and must participate in the building program in order to qualify for a new house.

The local organization became affiliated in September 1991 under the leadership of Dr. Charles Payne. The original board of directors were: Don Murph, Vickey Haynes, Bob Renner, Charles and Marietta Payne, and Dr. Bill Odom; Harvey Willis, Jonathan Washington, Catherine Orange, Roy Strother, Bobby McLain, and Linda Nelson.

Connie Allen was the first Habitat home owner. She moved into her home with her son Courtney on June 6, 1993. It is located at 401 College Street. The second Habitat home was built at 505 South Street in March 1994. The third house was completed in April 1997. The fourth house was completed in February 1999 and the fifth house was completed in March 2000. Two more partner families have occupied new homes and 4 more houses are nearing completion in 2001.

Charles Park became executive director of Habitat for Humanity in August 1999. Since that event occurred, the budget has grown from $35,000 to $250,000 per year. Under the leadership of the board and the new executive director the organization has expanded its service into all of Webster Parish. The goal is to build 4 or 5 houses during the next two years. The long range goal is to build more than 4 or 5 houses in the following years. Funds for the projects come from the Habitat Walk fund raiser, private donations, and contributions from organizations and businesses.

Lions Club. Lions International was founded in 1917 by Melvin Jones in Chicago, Illinois. Today the organization has 1.4 million members in 43,000 clubs in 185 countries. The Springhill Lions Club was founded December 14, 1927, as a local club affiliated with Lions International.

The motto of the club is "We Serve." This motto is expressed through numerous projects including, but not limited to, the distribution of eyeglasses to the people of Central and South America, Toys for Tots at Christmas time, co-sponsor of the

Lumberjack Festival, financial support for Habitat for Humanity, and a $500 scholarship for a graduating senior at Springhill High School; and the club also sponsors the annual football banquet. One of the major projects is financial support for the Louisiana Eye Foundation Research Center.

The emblem for the Lions Club was adopted in 1919. The gold "L" on a field of purple is flanked on each side by the head of a lion, one facing the past and one facing the future. Colors are purple and gold. Purple stands for loyalty to friends and integrity of mind and heart. Gold stands for sincerity of purpose, liberality in judgment, purity in life, and generosity in mind, heart, and purpose to humanity.

Mr. Wilbur Wilson, from Plain Dealing, Louisiana, and a member of the Springhill Lions Club, is a very important Lion. He worked with Melvin Jones in Chicago to develop Lions International and he served as the Executive Secretary of Lions International for many years.[7]

Lumberjack Festival Association. The first annual Lumberjack Festival was sponsored in 1984 by the Springhill-Cullen Chamber of Commerce. The first festival was a success so the Chamber Board decided it would be better to have a separate organization sponsor the event. As a result the Lumberjack Association was formed. The first fifteen directors were appointed by the Chamber Board of Directors.

The purpose of the association is to plan and promote the annual Lumberjack Festival, to publicize the good things about the Springhill area, and to assist in its economic and social growth. It does not give special recognition to any individual or organization, but it makes a strong effort to do what is best for the entire population. Cliff Carter is the new president.

Masonic Lodge. The Masonic Lodge is a fraternal organization that was founded on February 24, 1911, under the leadership of Brother Lofton. At the first meeting the group elected the following officers: J. L. Strickland, Worshipful Master; J. D. Stephens, Senior Warden; R. L. Ensey, Junior Warden; N. B. Smith, Secretary; W. M. Morris, Treasurer; S. B. Dickey, Senior Deacon; M. L. Browning, Junior Deacon; and Andrew Dees, Tyler.

In addition to electing officers at the first meeting, the group also made three decisions: first, they agreed to adopt the Grand Lodge; second, they agreed to meet on Saturday at 8:00 P.M. before the third Sunday; and third, they agreed to pay the Tyler 50 cents for each meeting.

The new Lodge met for approximately two months in the Woodmen of the World Building on Butler Street. On August 19, 1911, a building committee was appointed to construct a new wooden building. The second building was constructed alongside the first building. They were located where the present Citizens Bank and Trust Company now stands. The third building was a two-story structure located in the middle of the block on Main Street near the present-day Boucher and Slack Insurance office. The Lodge used the top floor and rented the bottom floor to a café owned by Mr. O'Bier. The Lodge moved to a fourth building located next to the present Louisiana Drivers License office. Meetings were held in this building for a considerable length of time. The Order of the Eastern Star was organized on June 13, 1916. They also met in the building. The fifth building is being constructed in 2001 on the southeast corner of Main and Church Streets.

The purpose of the Masonic Lodge is "to take good men and make them better."

A second purpose is charitable. The primary focus of their charity is five dyslexia schools in Louisiana. The local lodge meets on the 1st and 3rd Thursdays at 7:30 P.M.[8]

Quarterback Club. The Springhill Quarterback Club was organized over fifty years ago to serve the Springhill High School football team. It fulfills this purpose through financial gifts and through generating enthusiasm for the team. Financial support comes from a variety of sources: fund raisers, membership sales, stadium sign sales, donations, scrimmage games, public address advertising at games, Lumberjack merchandise sold through the Jack Shack, and through Lumberjack flag sales.

The money raised goes toward supporting the Lumberjacks in various ways: assisting with pep rallies, promoting off-campus spirit events, furnishing pre-game meals for players and coaches, and purchasing sophomore jackets; assisting with financial support for football equipment, assisting with improvements to the stadium and weight room, providing team travel shirts, promoting Homecoming activities, and promoting the Lumberjack Festival.

Expensive items are purchased when funds are available for them. They include coach's headphones, varsity travel bags, blocking sleds, weight room equipment, and 4-wheelers. A major lighting project was completed in 2000. New stadium lights were purchased and installed at Baucum-Farrar Stadium. Parents of players are encouraged to join the Quarterback Club to support their sons.[9]

Rotary Club. Rotary International was founded in 1905, the world's first service organization. It has grown into a global network of 1.2 million members in 29,000 clubs in 160 countries. Members, whose motto is "Service Above Self," meet weekly to hear speakers and to plan programs. Club membership represents a cross section of local business and professional leaders who provide humanitarian services, encourage high ethical standards, and help build goodwill and peace in the world.

Through the Rotary Foundation $90 million is distributed for international scholarships, cultural exchanges, and humanitarian projects that improve the quality of life for millions of people. The eradication of polio is a top priority which requires the immunization of every child under five years of age in the world.

The Springhill Rotary Club was organized in 1957 under the leadership of Emery Diamond. Current officers are William Smith, President and Kay Tatum, Secretary-Treasurer.

The club participates in Rotary International projects and in local projects planned by the club. These include a $500 scholarship for a graduating senior, a donation to Frank Anthony Park, and funds for two students to attend Camp Ryla for leadership training.

Senior Friends. The local chapter of Senior Friends has over 1500 members sponsored by the Springhill Medical Center. It is affiliated with the National Association of Senior Friends which was founded in 1986 for the benefit of senior Americans. The mission is to provide healthy options for happier living for adults age 50 and over.

The local chapter members participate in numerous programs including Senior Friends Supper Club, parlor games, planned trips, National Health and Fitness Day, and free health screenings; bingo, breakfast fellowships, weekly exercise and dance activities, and guest speakers on topics of interest. The most important benefit is the opportunity to be with others in a healthy and active environment.[10]

Springhill Art League. The Springhill Art League was organized August 6, 1969, in the home of Erma Boucher. There were 19 women present. Originally it was named "Canvass Pals," but the group changed its name to the "Springhill Art League" and was incorporated October 19, 1976. Laphelia Middlebrook of Texarkana encouraged the group to organize into a club. The first officers were Hazel Harris, President; Erma Boucher, Vice-President; Vivian Blackwell, Secretary; and Esther Hill, Treasurer.

Purposes of the League are: to maintain and increase interest in art, to create opportunity for sharing knowledge and implement the talent and skill of members, to provide opportunity for instruction by arranging classes and workshops, and to provide for exhibit of arts and crafts.

Nationally known artists have taught seminars for the Art League. John Birdsong, Shelia Parsons, and Sam Tuminello are among those teachers. Workshops in water color, oil, acrylic, pastel, drawing and pen and ink have been offered to members.

The Art League has held exhibits in various places including, but not limited to, Pierremont Mall in Shreveport, Minden Civic Center, Fountain View Nursing Home in Springhill, Springhill Medical Center, and the Springhill Art League building and gallery.

The Art League sponsors the Northwest Louisiana Arts and Crafts Show and Sale that began as a fund raiser 26 years ago. They purchased a building at 222 North Main Street with these funds and those from other fund raisers. The beautiful and colorful gallery is used for meetings, workshops and exhibits. The Springhill Art League has been a catalyst and promoter for visual arts in the community.[11]

Springhill Riding Club and Rodeo. A group of 13 men met at the community house on July 1, 1952, and agreed to spread the word that all persons interested in a riding club should meet on July 6, 1952. These men were: Aubrey Mouser, Henry Smith, George Taylor, Reese Ferguson, and Paul Slack; Leo Thompson, Billy Twitty, Kermit Cone, J. C. Hilburn, and Lewis Barnard; J. C. Brown, Ernest Pickard, and Henry Hornsby.

Sixty-seven riders were present at the July 6 meeting. Billy Twitty was elected temporary president; Retha Cone, secretary; George Taylor, vice-president; and Adene Mouser, treasurer. In a later July meeting the group decided on black and gold as colors and a gold star on a black riding shirt to represent the Springhill Riding Club. On August 12 the group decided to order a club flag with 13 stars to represent the 13 men who were at the original meeting. The first club visit to a rodeo was July 22 in Magnolia, Arkansas.

On March 6, 1953, the club decided to purchase land for a rodeo arena. This was accomplished on March 19, 1953. In April the board decided to issue bonds for sale to the stockholders for the arena. On May 4, 1953, the club received papers officially designating it as the Springhill Riding Club, Inc. Construction of the 6,000-seat arena began on May 18, 1953.

On July 27, 1953, the club invited Tommy Steiner to provide the stock for the rodeo. The first rodeo held in Springhill was sponsored by the Springhill Rodeo Association in the new arena on August 20-22, 1953.

The present stock contractor for the rodeo is Simmons Rodeo Co. The present officers of the Riding Club are Willie "Butch" Lynd, president; Waymon Slack and B. J. Benefield, vice-presidents; and Eric Simmons, secretary-treasurer.

Springhill Senior Center. The Webster Council on Aging began Senior Center work in Springhill in 1970. The group first called themselves "55 Plus Club." The first facility used for a meeting place was the old Court Room next to the Springhill Police Station. They met once a month for entertainment, games, crafts, trips, and speakers. As interest grew, meetings increased to once a week.

Ed Kenyon donated a house to the city in 1971 that was to be used for a Senior Citizen Center. The group began meeting five days per week to exchange ideas, express creativity, gain education, receive encouragement for continued involvement in the community, and to enjoy a noon meal. As the Senior Center continued to grow a 12 x 30 room was added to the Center to accommodate the group and expand their activities.

Today the Senior Center is open Monday—Friday from 8:00 A.M. to 4:00 P.M. Directors of the Center have been Billie Hair (1970-71), Olene Sawyer (1971-1981), Geraldine Jackson (1981-1991), and Joyce Friday Shirley (1991-present).

The Webster Council on Aging provides a variety of services for the elderly through the Springhill Senior Center. They include Information and Assistance, Outreach, Assessment and Screening, Legal Assistance, Congregate Meals, Home-delivered Meals, Homemaker Services, Recreation, Ombudsman to Nursing Homes, Wellness, Elder Waiver, and Medicare Enrollment Center.[12]

Woodchoppers. Springhill Woodchoppers is an organization of approximately 40 people who go to campsites in motor homes once a month for a three-day campout. Their camp sites, food service, and group activities are planned one year in advance for these monthly trips.

The local chapter is a member of the national Good Sam Club and the State of Louisiana Good Sam Club.

Many of the local club members have participated in the development of Frank Anthony Park and in the remodeling of the old A & P building. These efforts will provide facilities for campers to stay overnight in motor homes and for campers and local civic groups to have access to a meeting place. Fannie Moore is president of the Woodchoppers Club.

The old A&P Grocery building is a public meeting place and is being converted into the Community Activities Center.

Left: *Entrance to the recreational park built on the site of the Springhill sawmill.*

Below: *The barbecue pavilion in Frank Anthony Park.*

Right: *Masonic Lodge on north Main.*

Below: *Habitat for Humanity hosts its first collegiate volunteers from Massachusetts. Charles Payne on ladder, Dale Thomas third from right.*

Top: *New Masonic Lodge building at corner of Main and Church.*

Middle: *Senior Friends Association Building on Medical Drive.*

Below: *Display in the front windows of Springhill Art League on Main.*

Chapter 11

Great Pines–Recreational Growth

Springhill is a town that enjoys recreation—fishing, hunting, water sports, ball teams, rodeos, parades, and festivals. All ages participate in recreation which is defined as a "pastime, diversion, exercise, or other resource affording relaxation and enjoyment."[1]

Northwest Louisiana has been known as "Hunter's Paradise" since 1824 when fires cleared the underbrush and thick forest that allowed wild game to migrate into the area. Plentiful water in lakes, rivers, and creeks is available for fishing, water sports, and camping. The Louisiana State Tourism Commission has identified the entire North Louisiana area as "Sportsman's Paradise."

Championship football and basketball have been played in Springhill for many years. Enthusiastic fans support these winning teams. Ducks, quail, and doves are abundant for hunters. Some of the finest bass, catfish, and crappie fishing is available in lakes, creeks, and rivers located within short driving distances from downtown Springhill. A rodeo is held every year. Three parades during the year provide entertainment: Mardi Gras Festival parade, Lumberjack Festival parade, and the Christmas season parade.

Five major events influence recreation in Springhill. First, the State of Louisiana Athletic Association accepted Springhill into its membership in 1927. The event created competition in high school athletics among neighboring cities in organized district or league sports. This recognition of the athletic program of Springhill High School advanced the development of sports immensely.

The second event was the emergence of organized baseball in Springhill. It came first through the men's baseball team sponsored by International Paper Company. A ball park was constructed in Pine Hill. The team played semi-pro style baseball in the late 1940s and early 1950s. It was disbanded in the late 1950s and the stadium was torn down.

The third event was influenced by the high school boy's baseball team. An American Legion baseball team was active in the early 1950s. This was followed by

Little League baseball and Dixie League baseball for boys in 1958 and 1959. These teams encouraged youth to play baseball in organized leagues.

The fourth event was the construction of Lake Erling in 1956. A 7,000-acre lake was built by International Paper Company a few miles north of the Arkansas state line in Lafayette County. It opened up camping, fishing, water sports, and family lake homes for the citizens of Northwest Louisiana and Southwest Arkansas. The lake was named after Erling Riis, chief engineer for the construction of the paper mill.

The fifth event was John David Crow's winning of the Heisman Trophy in 1957. He was recognized as the best collegiate football player in America. John David's superb athletic abilities created enthusiasm for four years during athletic competition at Springhill High School. His greatness as a football player was rewarded with the game's highest honor while he played halfback at Texas A&M University. This inspired the youth of the nation, and specifically the youth of Springhill, to play with dedication and discipline.

Baseball for men. International Paper Company sponsored a men's baseball team in Springhill in the late 1940s and early 1950s. The team was called the Springhill Sports. They disbanded a few years and emerged again in 1956. The team played in a ball park in Pine Hill near the paper mill.

Smoke would often fill the stadium, thus making it difficult for both spectators and players.

Managers for the baseball club were Ed Harris, Robert Bowles, Ed Thigpen, and Calvin Craig. The team disbanded in the late 1950s and the stadium was demolished.

One of the umpires working behind the plate was Mr. Jeter. He was short of stature so J. T. "Sleepy" Haynes gave him the name "Squatlow." During the baseball game it was a popular and fun thing to tease the umpire about his ball and strike calls. The fans would say, "Squatlow, get some glasses," or "Get up higher so you can see the strikes." Even Mr. Jeter seemed to enjoy the fun.

The food concession at the ball park was operated by Howard Smith. He employed school boys to sell Cokes and peanuts in the ball park. The boys received one cent for each Coke sold.

Three players had unique experiences during their baseball careers. Max Alvis, who played third baseman for the Springhill Sports, played later for the Cleveland Indians. Calvin Craig played shortstop and later managed the ball club in 1956. Melvin Lee Pig from Fair Park High School wanted to remain anonymous so he changed his name to Melvin Lee for his baseball name.[2]

Baseball for youth. "Batter up" is the cry heard on baseball diamonds of public schools in Springhill. Today the youth baseball program is called "Select Baseball" signifying that teams play for competition, but they do not participate in league play. This is a change from earlier youth baseball in the city.

American Legion baseball was played by high school boys from the early 1950s through the 1960s. Dixie Youth League baseball was played by boys nine years to eleven years of age. The Dixie Major League was for boys 12-15 years of age. After playing in the Dixie League, the boys played either on the high school team or the American Legion team. Today high school boys play baseball in the Select Baseball competition.

J. A. "Junior" Sanders, Mac Pace, Bill Mosely, Ed Olive, and Roy Nichols provided leadership as coaches and league officers for the American Legion baseball, Dixie League baseball, and Select Baseball from the early 1950s until today.[3]

Basketball. Coach William "Billy" Baucum was the only coach to win a state championship in both football and basketball. His 1952 Lumberjack football team beat Reserve High School to win the state championship. The 1953 Lumberjack basketball team won the state championship in Louisiana.

The girls basketball team won the state championship in 1955. Louise Wardlaw Lewis was the coach of the championship Lumberjills. The 1964 Lumberjills played in the Sweet Sixteen state finals, but lost the game. Ophelia Carrol was the coach.[4]

During other years the Lumberjack football teams and basketball teams have won district championships. Many citizens are enthusiastic followers of these athletic teams. When they have played for the championship in New Orleans, hundreds of fans make the trip to "root for their boys or girls."

Camping. Camping by Springhill residents is accomplished in three forms: individually, families, and groups. Hunters and fishermen often stay overnight to pursue their quarry. This form of camping is considered "roughing it" because the conditions are often primitive in the woods.

Families camp in motor homes or tents in campsites. They enjoy water sports, fishing, swimming or simply the great outdoors. Parents have discovered this to be a good way to know and train their children.

The third form of camping in the Springhill area is through organized groups such as the Woodchoppers. They are comprised of a group of 40 people who go to campsites once a month for three days camping in their motor or travel trailers. Frank Anthony City Park provides hook-ups for out of town guests to bring their recreational vehicles into the city for overnight camping.

Football. Football teams have a winning tradition in Springhill. The 1948 team led by Junior Turner won many games within the district and won a reputation as one of the best teams ever at Springhill High School. The 1952 football team led by John David Crow won the state championship by defeating Reserve High School in the state finals 20–13. The 1985 Lumberjacks played Homer in the Super Dome in New Orleans. They defeated the Pelicans 19–14 to win the state championship. The 1987 team lost to Jonesboro in the state finals in the Super Dome.

The contemporary winning tradition is attributed to Coach Billy Baucum who made an inspirational speech to the student body at the beginning of school in 1950. He illustrated the speech with World War II experiences he encountered in the Normandy invasion. Team effort, courage, and personal discipline were ingredients mixed with the belief that this is the best class in the best city in the best state in the best country in the world. It was an attitude adjustment from a depressed student body to one of enthusiasm with the winning spirit. Attorney Jack Montgomery, an alumnus of Springhill High School and its football team, presented this story to the football team at the annual football banquet in 2001. He said, "Many believe this was the turning point in 1950 that created a winning attitude and began a winning tradition in the student body."[5]

Some head coaches for football teams through the years were Ray Hudson, — Clabaugh, — Culberson, Clayton Cornish "Cracker" Brown, E. W. Brown, Ben

Clements, Butler Miers, Billy Baucum, Travis Farrar, Wilbert Flanigan, and Benny Reeder the current coach.

Gambling. Gambling is a controversial activity in Springhill. When the Relay Station truck stop, restaurant, and video poker room was announced for Springhill in 1998, concerned citizens petitioned the City Council to vote to keep gambling out of town. The vote was split in favor of the Relay Station. Travelers through town and some local citizens visit the video poker gambling machines in the Relay Station. This is their form of recreation.

Lake Erling. In 1956 International Paper Company built a 7,000 acre lake in Southwest Arkansas to provide a constant water supply for the paper mill. The lake was named after Erling Riis, chief engineer during the construction of the paper mill in 1937.[6]

Bass fishing, crappie fishing, catfish fishing, skiing, boating, and family outings have become popular recreational activities on the lake. It is a vital resource providing relaxation and enjoyment for the citizens of Springhill. Many families have built lake homes on Lake Erling.

Parades. Springhill residents enjoy the entertainment provided by four parades each year. The Mardi Gras parade occurs in February. It ushers in the religious season of Lent for many Christians. A king and queen are chosen for the celebration. A grand marshal of the parade is selected and rides on a float in the parade. One of the exciting things about this Main Street-sponsored parade is the throwing of the beads from the floats. "Throw me some, Mister" is often heard from the crowds. Jan Willis, Director of the Main Street Program, is the coordinator of this parade.

The Rodeo parade is held each August in conjunction with the week-long PRCA Rodeo in Springhill. The Rodeo Association is responsible for producing this entertainment. Cowboys, horses, and floats go up Main Street in parade fashion to publicize the rodeo. Butch Lynd is president of the Rodeo Association.

The Lumberjack Festival is a major event in Springhill each October. The festival is designed to promote Springhill and the surrounding area, its businesses and its people. Since 1983 there has been an annual Lumberjack Festival and parade to kick-off the weekend event. Ann Sanders has served recently as the Lumberjack Festival Association president. Cliff Carter is the current president (2001).

The Christmas parade officially opens the Christmas season in Springhill. It is held in December. The Chamber of Commerce is responsible for the production of this parade. Harriet Carter, the Chamber of Commerce Executive Secretary, is the project director for the parade.

Parlor games. Bridge is a popular pastime in Springhill. Several ladies clubs are active. Mrs. Avis Garrison, former owner of the *Springhill Press and News Journal,* taught the game of bridge to passengers on three Caribbean Sea cruises. She continues to play bridge during her retirement years.

Rodeo. The rodeo was begun in 1953 when a group of men purchased 14 acres on Machen Drive to build a rodeo arena. The arena was completed in 1953. John Wallace, Adene Mouser, Butch Lynd, and Eric Simmons have led the program from its infancy to its maturity. A rodeo parade is held in August each year. Eric Simmons supplies stock for the cowboys.[7]

Sportsman's Paradise. In 1824 Northwest Louisiana was a wilderness tangled with vines and thick underbrush. Transportation was accomplished on foot or on horseback along old Indian trails. At this time fires that had been started accidentally by hunters began to spread through the wilderness. As the underbrush burned, wild game and fowl began to enter the terrain. Especially deer and turkey entered the area. In a few years Northwest Louisiana became known as "Hunter's Paradise." [8]

Hunters enjoy hunting deer, turkey, raccoon, ducks, squirrel, quail and doves. Fishermen spend hours on lakes and creeks in pursuit of bass, crappie, and catfish. Campers spend days and nights camping in the beautiful forest and on the serene lakes near Springhill. Lakes are filled with skiers, swimmers, and fishermen. Citizens of Springhill live in the midst of a true "Sportsman's Paradise."

Boy Scout Hut in City Park on an icy day.

Lumberjack Lanes bowling alley.

Baseball field neighboring I.P. and south of Pine Hill Subdivision.

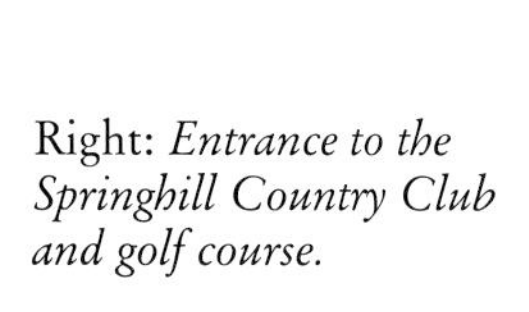

Right: *Entrance to the Springhill Country Club and golf course.*

Below: *Springhill Country Club golf tournament ca. 1951.*

Lake Erling's Percy Cobb Dam opening with crowd in attendance.

Springhill Rodeo Arena.

Circle "S" Riding Club.

Mayor Ed Shultz and "Babe" Sears breaking ground for the east side swimming pool about 1948. (Swan Clinic in background)

West side swimming pool.

East side swimming pool.

1954 Class A Louisiana State Basketball Champions, coached by Billy Baucum.

Baucum-Farrar Lumberjack Stadium on Hwy. 371 south of junior and senior high schools.

1952 Class A Louisiana State Football Champions, coached by Billy Baucum.

Gary Bonner and Noel Custer carry John David Crow off the field after Springhill tied Byrd 20-20 in 1953. Billy Chase looks on.

S. H. S. coach Billy Baucum, John David Crow and Texas A & M coach Bear Bryant on John David Crow Day.

1955 Class A Louisiana State Basketball Champions, the Lumberjills, coached by Louise Wardlaw Lewis. Barbara Alley, Frances McGowan, Johnnie Riggs, Ladelle Edwards, Alicia Williams, Barbara McMahen.

1985 Class AA Louisiana State Football Champions, coached by Travis Farrar.

Chapter 12

Great Pines–Religious Growth

North Louisiana History

South Louisiana Acadians and French practiced their faith in the Roman Catholic Church. North Louisiana was settled by Anglo-Saxon families from Middle Tennessee and the Southeastern United States. They practiced their faith in the Baptist and Methodist Churches.

North Louisiana settlers brought established religious denominations to the region. They were conservatives with a two-fold theological heritage—that which was brought to America from Europe by the colonists and that which came from theological divisions and controversies which emerged in the United States. A basic characteristic of North Louisiana frontier religion was its Protestant and Anglo-Saxon antipathy toward Roman Catholicism.[1]

Baptists and Methodists were the most numerous Christians in North Louisiana. Revivalism, emotionalism, hymn singing, and lack of ministerial education characterized both groups during their early history. The hill parishes of the northern part of Louisiana were less an extension of Europe than those of South Louisiana. They represented the westward movement of older Southern states. Pioneer religious leaders sought to reproduce churches and organizations that were similar to those they left behind both in theological dogma and ecclesiastical organization. Emory Stevens Burke, Methodist church historian, noted that the journals of pioneer ministers were written about "people in new places yearning for the old faith."[2]

These early settlers not only had a deep conservatism and a perpetuation of established ways that American colonists had transplanted from Europe, but they carried with them also theological controversies and divisions which had emerged in other American states. The anti-missionary group among Baptists, the Anglo-Saxon fear and mistrust of Roman Catholicism, and the rejection of the ecumenical spirit within Protestantism were specific areas of disagreement. As elsewhere in the South, the slavery issue divided the Baptists, Methodists, and Presbyterians.

Camp meetings were popular among Baptists and Methodists. Religion was practiced but the organization of churches was slow to develop. Old school Presbyterianism was anti-revivalism. Theological and ecclesiastical differences, such as baptism and church government, divided denominations more than Calvinism, the bedrock of theological understanding among Baptists and Presbyterians.[3]

It was in the midst of this conservative theological pattern and in the midst of denominational controversies among Baptists and Methodists churches in the nation that new churches began to develop in Northwest Louisiana. The histories of Springhill churches reflect that religious environment.

Church History Vignettes[4]

Central Baptist Church. The church was organized in 1922 following a revival conducted by Evangelist C. P. Roney of Shreveport. A difference of opinion about convention literature and about the method of supporting missionaries caused 40 members of First Baptist Church to pass a resolution to organize a Southern Baptist Church. Rev. J. P. Durham was called as the first pastor of the church which was soon named Murray Memorial Baptist Church. In 1939 it became Central Baptist Church.

When Rev. Luther Fortenberry came to the church as pastor in 1946, the members built a new education building and sanctuary. The front of the sanctuary was remodeled in 1971 under the leadership of Rev. Kenneth Everett. Other renovation projects have occurred through the years.

In 1957 property was purchased on Williams Street for a mission. In 1960 the mission was organized as a self-supporting church and named Temple Baptist Church.

During many pastorates various programs have been initiated by the church: library, senior adult ministries, youth ministries, music programs, Child Development Center, Mothers Day Out, and Special Ministries. In 2000 a television ministry was begun under the leadership of Dr. Ron Harvey, the present pastor.

The church has served the Southern Baptist Convention faithfully through generous gifts to missionary and benevolence causes. The congregation has been a strong spiritual influence in Springhill while serving the people of the community under the theme, "In the heart of Springhill with Springhill at heart."

In their 50th anniversary *Chronicle of Christian Growth 1922–1972* the history of the church was concluded with these words: "It is the purpose of this Great Church to maintain a house where worship of Almighty God is central, with the Lord Jesus our Master, the Bible our Creed, the world our Field, the Great Commission our Charter, and Faithfulness our duty."

Church of Christ. The Springhill Church of Christ was founded about 1941 when Christians began to meet in various homes. Mable Beavers was one of these persons. Preachers from Haynesville led worship services on Sunday afternoon. Two of them who preached to the congregation were Carrol Bailey and S. C. Kinningham. Brother C. D. Crouch was the first "located preacher" to serve the congregation. He served from 1943 until January 1945. He was supported by A. M. Burton, a Nashville, Tennessee, insurance executive.

A meeting house was completed in 1944 at 405 Butler Street from materials that were salvaged from a nightclub called "Edgewood Club." In 1945 a lot that joined the meeting house property was purchased by the congregation. In 1950 the congregation built a house for the preacher. During the pastorate of Dorice Mitchell (1959–1961) the preacher's residence was moved to Mill Pond Road and a new meeting house on Butler Street was constructed. In 1974 the preacher's residence on Mill Pond Road was sold and a new residence was purchased on Janice Drive. The church building has remained at the Butler Street location throughout the years.

During the summer of 1970 Aubrey Miller led in the establishment of another congregation that met on Main Street. The congregation split to establish a congregation called North Arkansas Street Church of Christ.

In November 1970, A.M. Sanders, Alvin Powell, and B.B. Stanford were appointed as elders. Two previous attempts to appoint elders in 1961 and 1963 had failed. The church supported Carlos J. Valenzuela in the Philippines. The congregation also had a radio program on Sunday morning and a weekly newspaper article.

Larry Powell, a principal from Plain Dealing, and now an employee at Bossier Parish Community College, fills the pulpit on Sundays.

Dorcheat Acres Missionary Baptist Church. The church was organized in the summer of 1967 with 19 charter members and a new building on Airport Road. Rev. Ray Doster was pastor.

Prior to 1967 the organized group of 14 began meeting in the Old Community Building in Springhill under the direction and ministry of Rev. Arley Powers. Rev. Carl Budwah is the present pastor of Dorcheat Acres Missionary Baptist Church.

East Side Baptist Church. East Side Baptist Church was organized as Immanuel Baptist Church February 8, 1953, with twelve charter members. The first building was constructed by members and friends. A lot was purchased on Reynolds Street in March 1953. The original building now serves as a nursery, Sunday School rooms, and library.

Rev. H. S. McLaren conducted the first revival in August 1953 which resulted in 10 additions to the church. He was called as the first pastor. From 1953 to 1989 there have been 12 pastors of the church. Seven men from East Side Baptist have heeded God's call and surrendered to the ministry. Rev. Randy Wilson is the present pastor. The name of the church was changed to East Side Baptist Church on May 26, 1954. There is no record giving the reason for the name change.

First Assembly of God. In May 1945 Rev. A. C. Ayers, Sectional Presbyter, assisted a small group of believers to meet and begin a church in an old service station located on the corner of Eighth N. W. Street and Plain Dealing Road. The group purchased property, called Rev. M. L. Hall as the first pastor, and made long-range plans to purchase other property.

The church conducted its first business meeting on March 13, 1947. They incorporated in 1948 with 14 charter members. Next, the church purchased lots on the corner of 10th N.W. and Spring Streets. A new auditorium was constructed in 1954 on the corner of Reynolds and 9th Street, Southeast. The present pastor is Rev. G. R. Collins.

First Baptist Church. Workers and their families who came to Springhill in 1896 to work in the Bodcaw Lumber Company sawmill needed a church. The Baptists made arrangements to meet at the Spring Branch Baptist Church in nearby Columbia County, Arkansas. In 1902 Evangelist J. C. Vaughn conducted a brush arbor revival on 7th Street N. W. A decision was made to immediately organize a Baptist Church. The church called Rev. Vaughn as their first pastor.

In 1907 J. F. Giles, Manager of Pine Woods Lumber Company, donated one acre of land and all needed materials to the town for the purpose of constructing a church building to accommodate the Baptists and the Methodists as a place of worship. The two groups met in the union building until 1920. The Methodists offered to sell their half of the union building to the Baptists if the Baptists would move the building to a new location and allow the Methodists to retain the land. Baptists agreed, paid $500 for the building, and moved it to the present site of First Baptist Church. The land on which the building was placed was donated to the Missionary Church of Springhill on April 17, 1920. The church is located on land bought in 1860 by Monzingo and Neal. A new sanctuary was constructed in 1962. The present pastor is Ronald Morgan, a Springhill native.

Harrison Chapel. The church began in 1905 when Professor Rosenbore gathered a small congregation for Sunday School under a brush arbor. Land was donated by J. F. Giles to build a sanctuary. It was located at the present site of Branch Brothers Motor Company on Main Street.

Rev. Harrison Brown of Camden, Arkansas, organized the congregation and served as first pastor. The name Harrison Chapel reflects the honor given to Rev. Harrison Brown when his first name was chosen for the church. The name "chapel" refers to a place of prayer and worship.

Harrison Chapel was relocated when Pine Woods Lumber Company closed its mill. The area where the church was located on South Main Street was converted to commercial property, so the congregation moved the church to Patterson Street, its present location. In 1996 a new sanctuary was dedicated. In 2000 a note burning of the mortgage was held, making the church free of debt.

New Bethel African American Methodist Episcopal. According to records the church was organized "a few years before 1921" in a building on South Main Street. Rev. Jesse Hodge was the pastor. In 1927 the church building was destroyed by a storm. It was rebuilt by the generosity of three ladies and the gift of lumber by Mr. J. F. Giles. In 1941 the church moved to its present site on 800 5th Street N. W. in order to make room for business development on Main Street. Mr. R. A. "Buck" Smith and his son Robert Charles donated land for a new sanctuary. The church was rebuilt under the leadership of Rev. T. H. Brown and Dr. G. H. J. Thibodeaux.

A series of pastors has led the church to grow in numbers, ministries, and facilities. A new worship center was completed in 1974. In 1975 Rev. Jonathan Washington became pastor at New Bethel. Through his ministry many things have been accomplished. It was written by a member of New Bethel, "We would be contrary to God's will if we did not look back to those who are still with us. For they are the ones who saw it from a chicken shack to a temple for Christ."

North Arkansas Street Church of Christ. A group of people within the Church of Christ on Butler Street were influenced by the "Anti-Movement" which began in Highland Church of Christ in Abilene, Texas. They had a difference of opinion about certain issues within the church, so they felt led of God to begin a new congregation.

In March 1972 the North Arkansas Street Church of Christ held its first worship service. Rev. Aubrey Miller preached the first sermon. Earlier, in August 1970, a group of people purchased a tract of land from the Jesse Calvin Dooley estate. The minutes of the business meeting held February 24, 1971, record that the church began plans for a new church building. In August 1971 a contract was signed to construct a sanctuary for $40,000. The mortgage loan obtained from Springhill Bank and Trust Company was paid off in a few years. Seven preachers have served as pulpit ministers of the congregation since 1972. The present pastor is Tim Hisaw.

In their history the church historian wrote, "The North Arkansas Street Church of Christ is a New Testament church and is striving to do God's will. The future looks bright; with hope, faith, charity, and prayer, we believe it will become even brighter."

Sacred Heart Catholic Church. Catholics began settling in the Springhill area in 1936 when they came to work at the International Paper Company mill. Pastor Rev. Robert De Vriendt from Minden led worship services for the new arrivals. Rev. Vernon Bordelon served Catholics in Springhill after reassignment of Father De Vriendt. He said mass in the community center.

The acquisition of land and subsequent construction of a building was initiated by Rev. M. L. Plauche of Minden in 1976. The Catholic Extension Society provided money for this building and for a Parish activities building.

A full time pastor, Rev. Francis O. Couvillon, was appointed in July 1952. The rectory was not constructed until 1954, so Father Couvillon resided in the sacristy. Thirteen priests have served the church since its beginning. The present priest is Rev. Karl Daigle.

Sacred Heart Catholic Church celebrated its 40th anniversary as a Parish in June 1992. By decree of Bishop Friend in July 1993, the church became a quasi-parish (mission) of St. Paul Church in Minden. In June 1999 the community of Sacred Heart celebrated its 50th year as a vibrant Catholic presence in Springhill.

Springhill Christian Church. A group of concerned people met in Springhill on Saturday night, April 28, 1973, to establish a New Testament Christian Church. As a result of this meeting the Springhill Christian Church was organized with 20 charter members. The first service was held May 6, 1973, at the Axe Restaurant with attendance in the 20s.

Rev. Lewis Lyles, pastor of Westview Church in Shreveport, visited Springhill and preached for the new congregation until the church called its first pastor, Rev. Ralph Posey, of Clarksville, Georgia. The congregation was composed of 58 members when this event occurred.

The church met at the Axe Restaurant for 16 months until they moved into their new building located on 11th and Machen Streets in October 1974. A new wing was added to the building. The church began use of it in March 1977. A full schedule of activities for all ages was planned and a bus ministry was begun that same year. Through sacrificial giving by the congregation the debt was paid off in March 1977.

The church has been led by seven pastors. The current pastor is Rev. John J. Downs, a graduate of Springhill High School. In their church history it is written, "The Springhill Christian Church is very thankful to the Heavenly Father for granting us many victories in Jesus Christ."

Springhill Missionary Baptist Church. The church was organized on March 2, 1952, in the Springhill Community House with 17 members present. The charter was left open through a two-week revival for 21 more members to be received. This made a total of 38 charter members for the new church.

Rev. J. E. Hollingsworth was the first pastor. Mrs. John Brady was the first church clerk. Lester Norton was the first treasurer. The new congregation affiliated locally with the First Springhill Missionary Baptist Association, the Louisiana State Association of Missionary Baptists, and the American Baptist Association. Through these associations the church participates in the support of missions, benevolence, and Christian education.

In 1955 the church constructed a spacious and attractive facility at 102 Forest Street. The Building Committee was J. E. Brady, E. L. Clements, F. F. Eason, A. T. Boyer, and E. T. Hornbuckle. The church owns one-half of a city block with adequate facilities and parking. The church has been served by fifteen pastors through the years. The present pastor is Jerry Carter.

Springhill Presbyterian Church. On September 24, 1995, the congregation petitioned the Presbytery of the Pines to dissolve the Springhill Presbyterian Church, effective October 15, 1995. All assets of the church were transferred to the Presbytery of the Pines. The last worship service was held October 15, 1995, under the leadership of Rev. Jack M. Kennedy, pastor.

Fifty-five years and eleven months prior to its dissolution, on October 29, 1939, the Presbytery of the Red River, through its commission, organized the congregation. A charter was applied for and granted November 21, 1946.

Presbyterianism came to Springhill July 23, 1939, when the first service was held in the home of Mr. and Mrs. S. S. McGill. Six adults and two children were present. Palmer W. Deloteus, a candidate for the ministry, conducted worship services in Springhill in the early months of congregational life. He was sponsored by the Men's Bible Class of the Minden Presbyterian Church. The fruit of his work resulted in the granting of a constitution and the arrival of the first pastor, Rev. Alex W. Hunter on October 29, 1939.

Worship services for the congregation were conducted in the City Hall and State Theatre until a new building was constructed in 1940. The first service was held in the new building September 29, 1940. A church bell was donated to the congregation by the L & A Railroad in February 1953. It was placed in a belfry which had been given by other members of the church and personal friends in memory of Clyde C. Harper. The church became known as "the church that rings the bell." A beautiful new sanctuary was dedicated in February 1975.

The church has been led by eleven pastors during its 55 year history. Notable achievements include the construction of two buildings, enrollment and accomplishment in the Town and Country Church Achievement Program, construction of the Boy Scout Hut, and partial support for two Korean missionaries.

Southern Methodist Church. A concerned group, Mr. and Mrs. Darrell Willis, Mr. and Mrs. Robert Andrews, and Mr. and Mrs. Marshall Sanders, visited with Rev. Frank Beauchamp of Shreveport about starting a church in Springhill. The group began to meet in Marshall Sanders' home June 2, 1968; later they met in the Union Hall on Main Street.

The church was officially organized by Rev. Frank Beauchamp on July 14, 1968, with 16 charter members present. The new church purchased a piano for $125 and 50 "Favorite Hymns" songbooks.

A new sanctuary was begun with the purchase of 2.445 acres of land from International Paper Company August 9, 1969. Groundbreaking ceremonies were conducted on October 12, 1969. Construction was begun by Dodson Brothers Construction Company on October 24, 1969. The first worship service was held in the new building November 19, 1969. That building was sold and the congregation purchased the land and buildings of the former Presbyterian Church on April 26, 1996. The congregation has been led by ten pastors since 1968. Rev. John Youngblood is the current pastor (2000–).

Springhill United Methodist Church. The seeds of the Methodist Church were sown when Rev. R. M. Blocker started a church east of present-day Porterville in 1885. In 1895 Rev. R. W. Vaughn, a minister in the Methodist Episcopal Church, came to Blocker's Chapel as their new pastor. In 1895 he visited Springhill to conduct worship services for a small group of Methodists in the area. This was the first meeting of Methodists held in Springhill. In 1896 Rev. Vaughn returned to Springhill to start a Sunday School for these Methodist Christians. Services were held in a small building constructed in 1896 by Pine Woods Lumber Company. It was a building used by all the churches in the community.

In 1897 the Louisiana Conference appointed Rev. William Martin as the first full-time pastor. He quickly organized the first official Methodist Church in Springhill in that same year, October 1897.

The small building constructed in 1896 was destroyed on March 9, 1902, as a result of high winds. From 1902–1907 worship services were held in a building on the east end of present-day Coyle Street. This was the school building for Springhill which was used for worship services by all Christians in the community.

In 1907 a building was constructed by Pine Woods Lumber Company on the location of the present sanctuary. In 1918 the Methodist congregation had grown large enough to need a church of their own, so the building was sold to the Baptists for $500 when they agreed to move it to the present site of First Baptist Church. The Methodists built a new sanctuary in 1918 on land donated by Pine Woods Lumber Company. It was on the site of the present sanctuary. Additional rooms were added in 1931 and 1940, but the church building remained essentially the same until 1950.

A new Sunday School building was completed in 1951. Church growth continued, causing the congregation to begin building a new sanctuary in 1951. It was completed in 1953. On May 24, 1957, Bishop Paul E. Martin dedicated the new educational building and sanctuary.

Continual growth created a need for a youth center so the congregation built a youth center immediately west of the sanctuary. It was opened for use on March 31, 1963. A major renovation of the sanctuary occurred in 1994 to pro-vide a renewed and refreshed worshipful setting.

A brief history of the church written in the Commemorative Centennial Program reads,

> But the history of Springhill United Methodist Church is so much more than buildings and financial campaigns. Even though God does love for us to share our earthly resources for His church, the real life-blood of the Springhill United Methodist Church is its people. Throughout the past 100 years, the individuals and families who have worshiped God in our church have been dedicated, generous, creative, loving, and supportive. We have shared our glorious times and our sor-

rowful time . . . And through it all we have acknowledged that it is God our Father who has instilled in us His Holy Spirit which has led us and carried us through the first 100 years of the Springhill United Methodist Church.

Temple Baptist Church. This church was born out of a Backyard Vacation Bible School at 300 3rd Street S. W. in the Zimmerly house. Three women, Mrs. K. A. Kendricks, Mrs. Mattie Nesbitt, and Mrs. Betty Stevens taught the Bible to 25–30 children each Friday afternoon. The ministry grew and was moved to 103 Edgar Street where the first revival was held in the church. The church is now located on Williams Street.

Under the administrative direction of Raymond Jones, 15 men and women began Baptist Temple as a mission of Central Baptist Church. O. M. Slack gave the land for the mission which was developed under the leadership of Rev. Dean Elkins, pastor of Central Baptist Church. It was organized into a Southern Baptist Church on August 7, 1960, and the name was changed to Temple Baptist Church. Rev. Jimmy Brossette was called as pastor of the mission in 1959. He became pastor of the church when it was organized in 1960.

A second building and an auditorium were constructed in 1963. In 1983 the church constructed another new building which was used for Sunday School rooms. In 1992 the members of the congregation enlarged and renovated the auditorium, updated the baptistry, and installed a new sound system. The church has been led by ten pastors and nine interim-pastors. Rev. Wayne Reeves is the current pastor.

Trinity Worship Center. A small group of men and women began meeting in area homes for Bible study. The group was largely influenced by the charismatic movement, The Full Gospel Business Men's International Fellowship, and the teachings of Evelyn Morris Niles. A baptismal service was held in the home of Gay and Bob Moody on Lake Claiborne where 32 men, women, and children were baptized.

In 1975 the group was invited by Phil and Jon Hull to meet in the Springhill Christian Academy. It became officially incorporated as Trinity Chapel on February 28, 1984. The first pastor was Evan Henderson. He was followed by Mary and Tommy Brown who have served since 1984.

A sanctuary, fellowship hall, Sunday School rooms, and offices were completed in 1987. The church continued to grow which created a need to build a Family Life Center. Ministries expanded to include Love Thy Neighbor, a benevolent organization, Trinity House to minister to physically and sexually abused women, and special services for the community.

In 1994 the name was changed to Trinity Worship Center. It has become an interdenominational, international, interracial church following the Great Commission in Matthew 28 and giving the glory to God.

Walnut Road Missionary Baptist Church. Walnut Road Baptist Church began as a self-supporting mission in 1956 sponsored by the First Baptist Church of Cullen. The first worship service was conducted March 10, 1956, in a small building on the old Sykes Ferry/Springhill Road, known as Walnut Road. On December 7, 1956, the mission was organized into a New Testament Church with 36 charter members.

Rev. Jesse Perkins continued to serve as the first pastor until July 1957. In October 1957 the church endorsed Rev. Perkins as missionary to Blocker Baptist

Mission. That mission was organized into a church in 1964, at which time Walnut Road granted membership letters to 48 members to help establish the new church.

In 1956 the congregation constructed an auditorium next to the original building. The church expanded and refurbished the physical plant several times to meet the needs of a growing congregation. Walnut Road Baptist Church purchased three acres of property on Highway 157 from Central Baptist Church in September 1993. They voted to build a new sanctuary and Sunday School building in March 1994, had groundbreaking in 1998, and conducted the first worship service in the new location in July 2000. There have been eleven pastors of the church since 1956. Rev. Randal Murphy is the current pastor.

Records reveal that 500 persons have become new members of the church through baptism or transfer of membership from other Baptist churches between 1956–2001. Several men have been called out by God and licensed to the Gospel ministry.

Washington Church of God In Christ. The church began to take root in 1947 under the leadership of three pastors in that period of time, Elder S. D. Daniel, Elder A. D. Kirkpatrick, and Elder Burnice Bolden.

Elder A. D. Washington was called of God to minister in Springhill and to lead the congregation to build a house of worship. Prayer services were held in the Ray Quarters Community for the building program. Although there was opposition to the building, according to a written history, "Elder Washington continued to teach and preach the Word of God, Holiness or Hell." The building was completed in 1949.

Other pastors appointed were Elder Earl Harris (1980), Superintendent Stephen Bradley, Jr. (1980), and Elder Edward Williams (1998) The church history stated, "The goal is to create a church for the future, today."

Ecumenical Ethical Issues[5]

Churches in Springhill, Cullen, and Sarepta have co-operated on three occasions to oppose legalizing liquor by the drink, the sale of pornographic materials and drugs, and the opening of the Relay Station video poker gambling room.

During the 1950s Christians met at Central Baptist Church to plan a strategy to ban the legalized sale of alcoholic beverages in liquor by the drink form. People canvassed the area door to door to gather signed petitions against the proposed law. The election to ban alcoholic beverages in this form was successful at that time.

In 1976 a campaign was organized to prevent the sale of pornographic materials and drugs in the area. A North Louisiana Moral and Civic Foundation was organized as the vehicle to fight the sale of these materials and drugs. Before the election was held in 1978, an open meeting for the community was conducted at Springhill High School to inform the citizens about the problem so they could become informed voters.

The third campaign was conducted in 1999 against the opening of the Relay Station video poker gambling room. Concerned citizens attended the City Council meeting to voice their protest against the legislation that would allow video poker to be played within the city limits.

Central Baptist Church, 1947.

Central Baptist Church new building, 1948.

Central Baptist Church, 2001.

Springhill Presbyterian Church, ca. 1954, before brick building built.

Southern Methodist Church, 2001, (formerly Springhill Presbyterian Church).

Springhill United Methodist Church, 2001.

First Methodist Church, 1948.

Church of Christ on Butler St.

First Baptist Church, 1948.

First Baptist Church, 2001.

New Bethel A.M.C. Church.

Church of Christ on N. Arkansas St.

Top: *Washington Church of God in Christ.*

Center: *Trinity Worship Center.*

Bottom: *Springhill Christian Church.*

Harrison Chapel Baptist Church.

First Assembly of God.

Walnut Road Baptist Church.

Temple Baptist Church.

East Side Missionary Baptist Church.

Sacred Heart Roman Catholic Church.

United Pentecostal Church.

Springhill Missionary Baptist Church.

PART FOUR

New Growth–

The Forest Renews Itself

1979-2002

Chapter 13

The Pines Fall–Economic Loss

Pine trees are grown as a cash crop. Small seedlings are planted in a nursery for the purpose of giving them an opportunity for early growth with the intention of replanting them on a tree farm. At some point they are sold, harvested, and transported to a mill to be cut into various sizes of lumber. This is the normal process.

The difficulty arises when pine beetles or diseases enter the growing trees and kill them. Natural disasters can destroy them in like manner. Economic loss will occur regardless of the reason for the tree-kill.

Economic systems of nations and cities act and react in a similar manner. Numerous factors combine to create a growing gross national product, positive quarterly earnings of companies, and a healthy bull market on the stock exchanges. If the bull market declines to become a bear market, severe economic loss will occur. The city of Springhill is affected by national economic trends and locally by business cycles. The city has experienced economic boom times and economic losses. It began as a small farming community that had tough economic times. When William Buchanan built his sawmill in 1896 the town began to have good economic times. It ceased to be a paper mill town in 1979. Local banks were founded and then closed over a period of years. The sawmill closed in 1972 and the paper mill closed its production division in 1979. These economic losses are examined in this chapter.

Sawmill policy. The "cut and move" policy of lumbermen in the north in the 1860s and 1870s was like a marching army across the northern half of the country from New York and Pennsylvania to the Great Lakes. There was no reforestation. The lumber companies cut all the good timber in an area, then moved the mill to another location. Economic loss was severe.[1]

William Buchanan had a "build and hold" policy with his mills, so he did not move them. Springhill was fortunate to have a stable economy with the sawmill for many years. It declined during the Great Depression in the year 1934 when Pine Woods Lumber Company closed its mill. Economic loss occurred when there was no industry and the bank closed. Futhermore, the national depessions hurt the city severely. During oral interviews of citizens all agreed that no one had any significant wealth in Springhill, yet the people of the area did not suffer from lack of food during the Great Depression. Most of them were farmers.

Earlier depressions. An economic depression began in 1873 and lasted six years. It grew out of over-extended credit, inflated currency, and the dispute over gold and silver as a basis for the economy. The monetary conflict finally precipitated a collapse in the entire economy.[2]

Early settlers who drifted into Northwest Louisiana after 1810 were self sufficient. They depended on the land for their livelihood. However, the fall of crop prices and the scarcity of loans hurt the farmers of Barefoot.

The depression of 1893 was caused by political fighting over gold and silver standards. As the silver surplus grew, its value deteriorated. Farmers in the Barefoot area were desperate to sell their timber in order to produce cotton. Buchanan took advantage of the situation and bought timberland from the farmers. He later built a sawmill and other amenities in the town. Farmers' poverty and despair were enormous.[3]

The great depression of 1929 began on October 24 when prices dropped rapidly on the New York Stock Exchange. Stockbrokers issued frantic calls for margin payments that spelled bankrupcy for thousands of small investors. Kate Butler described the effect of the depression on Springhill.[4] Farmers watched as their land worth $40 per acre decreased in value to $10 per acre. Cotton dropped to five cents per pound. President Roosevelt issued an order that every bank in the country be closed in March 1933. The Commercial Bank and Trust Company, after operating a year on a restrictive basis, closed for liquidation April 10, 1934.

A few months after Roosevelt's Bank Moratorium Directive was issued in 1933, Pine Woods Lumber Company suspended operations. Kate Butler wrote, "It was then that Springhill reached its lowest ebb. With its only industry gone, leaving hundreds without jobs, with the farm products glutting the market with no takers, with no banking facilities, Springhill was indeed struggling in the throes of its worst days."[5]

Paper mill closing 1979. The construction of the International Paper Company mill in 1937 created an industrial center with a booming economy until 1979. The five-point announcement by John Nevin on October 17, 1978, set in motion the closing of the four production machines at the mill. Approximately 1,000 people were affected.[6]

The company retained some facilities and built a wood products complex, but the era of the paper mill boom was over. The development of an industrial park attracted some small industries, but Springhill has never been the same. Economic loss occurred again.

Bank openings and closings 1916–2000. The founding and the closing of banks affected the economy between and in Springhill. The Bank of Springhill was established November 4, 1916 when businessmen on "The Hill" and in the community decided they needed a bank that was not under the influence and control of William Buchanan. The bank eventually closed.

The Commercial Bank and Trust Company had a brief life during the Great Depression. It was organized in 1933 and closed April 10, 1934.

Minden Bank and Trust Company established a depository in 1934, then transferred it to the People's Bank and Trust Company in 1935. It was converted to a branch bank in 1938 and returned to a depository in 1943. Although it did not cause a depression, the volatile banking situation in Springhill was inadequate for a stable economy.

Some progressive businessmen formulated plans for a locally owned and operated bank. They obtained a charter and opened Springhill Bank and Trust Company in April 1943. Springhill Bank and Trust Company operated as a locally owned bank until November 1988 when the Regions Bank Corporation purchased it.

Economic losses and depressions were not always the result of the founding, closing and selling of banks in Springhill, but the yo-yo environment of these financial institutions was a factor in the economic stability of Springhill.

Springhill oil boom 1951. Although oil was discovered in various cities and towns in Northwest Louisiana before it was discovered in Springhill, (Shreveport 1870, Homer 1918, Haynesville 1921, and Shongaloo 1921) a successful well was drilled in Springhill city limits in 1951 near the state line of Arkansas and Louisiana. Oil had been discovered in surrounding rural areas prior to this well: Dorcheat, Timothy Road, New Carterville, Walker's Creek, Spring Branch, International Paper Company ponds. Oil was discovered from the Red River to Haynesville along the state line.

The oil "bust" occurred in the mid-1980s.[7] When oil activity declined, some economic loss was created with investors, but it was not a depression for the entire community. Oil and gas activity has increased once again along the state line to the point that many land owners are receiving substantial checks from producing wells.

Great pines grow and fall. The brief description of economic conditions in Springhill in the preceeding paragraphs reminds citizens of the town that economic times are not always good, but neither are they always bad.

Louis Rukeyser, a popular television stock market analyst, captured the spirit of the challenge and the change in American economic life when he wrote during the 1983 crisis,

> America is by nature an optimistic society, but we are living through the pessimistic heyday. Let's examine the traditional ideas of America, including the right of freedom and chance to live better, and ask whether we can help the fading dream to revive. The individual's role is crucial, for in the end a better American economy must be the product of the nation's people, not its political leaders.[8]

During times of economic loss in Springhill, individual citizens have united to fight back, and each time they have won the battle.

Spectacular fire at the Anthony Forest Products Sawmill on the morning of September 23, 1974.

Chapter 14

New Seedlings–The New Springhill

Springhill is like a new thread woven into the fabric of the old cloth that produces a colorful masterpiece of tapestry. It retains the hues of the past as the new thread is mixed with the rich colors of the present.

The dividing line of Springhill's history was 1979, the year International Paper Company closed its four paper production machines. The economic boom which began in 1937 when the mill was constructed was over. The city changed dramatically.

Like the Phoenix of mythology, a new city arose from the ashes of the past and a new Springhill has developed. Much of the old has been retained, but there is much of the new in the city. The following comparison between the old and the new will assist us in understanding the continuing historical development of Springhill.

The Old and the New Springhill

Comparing the old and the new. Although the closing of the production division of International Paper Company was the primary reason for the drastic change in the city, it was not the only reason. The closing of the sawmill in 1972, the integration of schools in 1970, and the end of the oil boom in the mid-1980s significantly pressured the citizens to change their lifestyles.

The old Springhill was built around the sawmill and the papermill. The new Springhill is built around diverse industries. The old Springhill had separate schools for the blacks and whites. The new Springhill schools are integrated. The old Springhill played in Sportsman's Paradise by fishing and hunting. Recreation in the new Springhill has expanded into areas like the Art League and genealogy. The old Springhill emphasized high school football. The new Springhill also participates in the Lumberjack Festival, parades, and special events such as children's little leagues. Oil in the old Springhill was discovered by seismograph teams, but oil in the new Springhill is found with satellite scanners and computer generated maps. Once business relied on the typewriter. Now the computer is used for word processing and bookkeeping. Office supply stores dominated the past business supply needs. Now a computer store and an internet are important to the business man. The old Springhill was dominated by Baptists and Methodists. The new city has diverse religious denominations practicing their faith and others who choose not to participate in religion.

Dramatic change has come to the town. This change has produced stress among the citizens, but innovative, visionary, and courageous leaders have led the city into a new era. The work of building a new Springhill is still in progress, but the direction and goals have been set.

Rebuilding the city. John Gardner, former president of the Carnegie Foundation for the Advancement of Teaching, and advisor to the United States delegation to the United Nations, wrote an exceptional book entitled *Self Renewal*. He wrote, "A modern view of the process of growth, decay, and renewal must give due emphasis to both continuity and change in human institutions ... a sensible view of these matters sees an endless interweaving of continuity and change."[1]

Leaders have observed the decline in economics, changes in lifestyles, and the grief associated with these changes, and they have said, "Let this be a beginning, not an ending." They have led the town into diversity and modernity as a new Springhill has developed.

New businesses and institutions. Avalon Technology is a new type business in the city. In 1999 Citizens Bank sold the internet business to the new entrepreneurs who expanded the business to include computer repair, sales and training. New technology was added to the local business world.

The Northwest Louisiana Technical College was created to train new workers with new careers when International Paper Company closed production lines in 1979. The school closed after a brief time and re-opened in 1994 with an Industrial Maintenance Technology curriculum. State Representative Doerge and State Senator Foster Campbell assisted the school in funding for multi-craft training. It is budgeted through the Louisiana Community and Technical College Board of Supervisors. Industry demands students who attend and are graduated.[2]

The old Springhill had a library in progress. It was established as Webster Parish Library Branch in 1929. Mrs. S. R. Emmons helped plan the library opening and then served as the branch manager. Books were located in schools. Between 1930–1936 the library was located in the commissary of Pine Woods Lumber Company at the lot on the corner of Ensey and Giles Streets. Later the library was housed in the building formerly occupied by Russell's Pharmacy next door to Payne's Pressing Shop. On August 11, 1947, the Library Board of Control voted to lease the ground floor of the new Masonic building located at 231 Main Street where it remained until Mack Memorial Library was constructed in March 1995.

Mrs. Verna Strater served as summer manager in 1932. Mrs. J. C. Branch was appointed branch manager in June 1948 after serving as summer manager since 1936. Mrs. Branch retired on March 31, 1968. Mrs. Leland Allen assumed responsibilities as branch manager until she retired in 1995. Evelyn Simmons is the current branch manager. Ralph Ensey served on the first board. James Allen was on the board from 1960–1975. Eugene Eason is currently representing Springhill on the library board.

The children of Mr. and Mrs. Willie Mack, Gladys, Jeanne, Donald, Dorothy, and Mary, donated the former Mack store building for renovation to be used for the Willie and Mary A. Mack Memorial Library. It was dedicated in March 1995.[3]

Frank Anthony Park is a new development in Springhill. It is located on the site of the old sawmill in the downtown area. The acreage where Bodcaw Lumber Company, Pine Woods Lumber Company, Frost Lumber Company, and Anthony Forest Lumber Company mills were located is now cleared and used as a motor home park with facilities for over-night guests. Next to the park is the former A&P Grocery building which has been renovated for an activities center. Joe Curtis has been a driving force behind the development of the park. It is named in honor of Mr. Frank Anthony.

Innovative programs. The city of Springhill has initiated several new programs and projects to improve the lifestyle of the citizens. The Main Street Program is directed by Jan Willis. It began in 1998 for the purpose of beautifying Main Street and attracting shoppers to the historic district stores. The idea was proposed by Denny McMullan. Grants have been obtained to assist merchants in re-decorating the facades of their buildings. The most recent $8,000 grant is from Louisiana Main Street Program.

The Main Street program encourages merchants to renovate their store fronts with awnings and paint. The Mardi Gras parade and the Spring Festival on Main Street are events promoted by the Main Street Program.[4] One former resident said to the author, "I have been away from Springhill a long time. When I drove through town a few years ago, I wanted to turn around and leave the city. It had deteriorated so much I didn't want to look at it. But I came back to town in 2000 and found the storefronts attractive, renovated, and appealing. Someone has done a lot of work in the downtown area."

New city facilities. Plans for a new police station were drawn in April 2001. The building will be a 3000 square foot structure that will house 14 inmates and offices for the Police Department.

A new fire station is to be constructed on three acres of land at the end of Machen Drive on Percy Barnes Road. It will house the seven vehicles used to fight fires and for rescue operations. The chief and 13 volunteers will have facilities for office space and training.[5]

Beautification projects. The new Springhill has beautification projects that improve the appearance of the town. Mayor Johnny Herrington has given strong encouragement and leadership to the Beautification Committee. They have placed Christmas lights in the city park and along streets, and on the Civic Center, and the Municipal Building. They have encouraged the citizens to keep a clean town.

New pavement has been laid on Butler Street, Main Street and Highway 157 to Shongaloo. A Centennial clock has been proposed by the Mayor for the north end of Main Street as a part of the Centennial Celebration in February 2002. The goal of all these projects is to create an attractive city for citizens and visitors to enjoy.

Murals on buildings. Private businesses have added to the beauty and history of Springhill. Citizens Bank and Trust Company commissioned an artist to paint a beautiful mural on the east side of their building that depicts early timber industry scenes. The building across the street on the corner of Main and Highway 157 has an interesting and attractive mural around the theme of the old Coca Cola bottle. These murals have added beauty to the image of the new Springhill.[6]

Hall of Fame. Jesse Boucher and Woodrow Turner have developed a Hall of Fame to honor graduates of Springhill High School who have excelled in their professional careers or in athletics. Inductees are selected on the basis of citizen recommendations and committee evaluations.

Festivals. The first annual Lumberjack Festival was sponsored in 1984 by the Springhill—Cullen Chamber of Commerce. After its successful beginning the Chamber board placed the responsibility for the festival with a separate organization. As a result the Lumberjack Festival Association was born to plan and promote the annual event.

The purpose of the festival is to promote the city of Springhill, its business and industry, and its people. The major event of the festival is the timber sports competition. Booths located at the city park provide food service and arts and crafts displays. Artists, both local and out of town, perform at different times during the festival to entertain the public.

The October festival draws over 4000 people from the surrounding area during the one and one-half day event. International Paper Company, CenturyTel, Entergy, and Webster Parish Tourism Commission are the major supporters of the festival, and have played major roles in it since the beginning.[7]

Dreaming of the Future

Springhill has a rich heritage based upon the pine tree and upon the sawmills and paper mill that harvested it. The new Springhill has been characterized by a diversity of industry, an expansion of volunteerism, a modern lifestyle embracing the innovative technologies, and racial harmony.

The North Webster Industrial District, Chamber of Commerce, and business men are constantly seeking diversified industries that will locate in Springhill. They appreciate International Paper Company but in the future they do not want to depend upon one major industry for economic stability.

The shock wave caused by the sawmill closing in 1972 and the paper mill closing production machines in 1979 pressured the leadership of Springhill to begin re-thinking the future. Certainly a reality check was done in regard to the economic situation. Also, citizens began to dream about a new Springhill. The author encountered the following ideas about the future from conversations with citizens.

Volunteerism. When the paper mill ceased production, many outstanding community leaders who were employed at International Paper Company moved from Springhill. Other volunteers were needed to give leadership to church organizations, Chamber of Commerce events, city government activities, and organizational programs. Those responsible for recruiting these leaders have dreamed of an army of volunteers to plan, activate, and promote the many government, business, organizational, and civic activities of Springhill.

Museums. Others dream of a museum building filled with artifacts that represent the city's history. Genealogists could collect histories of local families and place them in the museum for study. Video tapes of oral interviews made during the planning phase of the Centennial Celebration could be stored, catalogued, and made available for viewing. The Historical Committee would advance the cause of local history in rapid fashion if the museum becomes a reality.[8]

Dinner theatre. Some citizens want to convert one of the unused Main Street buildings into a dinner theatre. It would be used for drama, children's productions, and adult-led dramatic plays. This, too, is part of the dream.[9]

New optimism. These dreams need to be fulfilled if renewal continues in the city. The history of Springhill is the story of optimism in the midst of change and adversity. There has been a rebirth of optimism following the loss of industry. Dreams of future projects and activities will occur if this optimism continues to be the prevailing feeling among citizens. One individual remarked when the paper mill closed its production machines, "We'll make it . . . we always have . . . if everybody will get out and work for it . . . I mean everybody in Springhill. We've got too many things going for us not to make it." This has been the historic attitude of Springhill citizens. If optimism continues to be the dominant force within her people, the dreams will come true.[10]

U.S. Post Office.

Smith's South Central.

Brookshire's.

North Webster Industrial Park entrance sign.

Trane Company.

Wal-Mart.

Soil Products Incorporated.

Electric Melting Services.

The Iron Works.

Springhill Motor Company on South Arkansas.

Citizens Bank and Trust.

Pine Plaza.

Bailey Mortuary.

Office Dimensions and the Industrial Park's rail terminal.

Chamber of Commerce (former depot).

Above: *South Main Mall.*

Right: *American Tri-State Underwriters.*

North Webster Industrial Park Office Complex.

PART FIVE

The Forest Matures

Chapter 15

Stress in the Forest–Difficult Times

Depression Times

"I remember when!" is a phrase often used by individuals who experienced the trauma of the Great Depression. Family members talk of "Hoover villages" and "soup kitchens." Unemployment was rampant and food was limited. The entire nation, in fact, the entire world, lived through the turbulent years 1929–1939 called The Great Depression. It was not an isolated event. Long-term influences bore down upon the nation's economic structure until the breaking point came in October 1929.

Influences on Depression. The Civil War had reduced the South to rubble. Out of a population of 5.5 million whites and 3.5 million blacks, 258,000 men lost their lives and 200,000 more were wounded. Cities were gutted, farms burned, livestock butchered, ferries and bridges destroyed, and railroads torn up. By March 1865, it took sixty-five dollars to buy one dollar in gold. Inflation was out of control. Battle-scarred areas had mile after mile of desolated land. Poverty was rampant.[1]

In Louisiana Dr. Sol A. Smith of Alexandria wrote to General Kirby Smith, "Louisiana lies mangled, rent, and palpitating in supreme agony of a ruined and trodden-down people."[2]

Previous Depressions. Stress was put upon the economy by former depressions. The Panic of 1873 grew out of over-extended credit, inflated currency, and the gold-silver dispute. The monetary conflict finally precipitated a collapse of the entire economy. The Depression of 1873 lasted six years. It hit Southern farmers hard. Prices of crops fell and loans became scarce. At that time Springhill was a little village of poor farmers in a place called Barefoot. Economic conditions hurt them severely.[3]

The Depression of 1893 was caused by political maneuvering over the gold and silver standards which created stockpiling of precious metals. The Treasury of the U.S. had 380 million silver dollars and 157 million ounces of gold bullion. As the mountain of silver grew, its value deteriorated. People began taking advantage of the gold option on their notes. President Grover Cleveland attempted to stop the gold drain by repealing the Bland-Allison Act in 1893. The economy continued to weaken. Farmers in the area of Springhill were desperate to sell their timber in order to produce cotton. Their poverty and despair were enormous.[4]

William Buchanan took advantage of the situation and bought timberland in Barefoot, Louisiana in 1894. This was the beginning of Bodcaw Lumber Company activity in the village which later became known as Springhill.

The Great Depression. The Great Depression began with the resounding collapse of the American economy in October 1929. Black Thursday, October 24th, saw more sell orders than buy orders. Prices dropped rapidly. Beleaguered brokers issued frantic calls for margin payments that spelled bankruptcy for thousands of small investors. By the close of the business day a record 13 million shares had changed hands.

On Tuesday, October 29th, the Dow Jones average fell 40 points. Sixteen million shares were sold cheaper than they were bought the week before the crash. The slide in the market continued into November, causing large and small investors to go bankrupt. Pessimism set the tone for business. Almost overnight the soaring spiral of rising prices reversed itself and became an equally dizzying spiral of decline. The result was economic disaster.[5]

Effect on Springhill. Kate Butler, respected citizen and wife of esteemed physician, Dr. Rupert Butler, described the effects of the Great Depression on Springhill when she wrote,

> Springhill, not unlike any other community in the United States, gradually began to feel the pinch of shrinking values. As time marched on, $40 land became $10 land with no takers; industry slowed up, cotton dropped to 5 cents per pound, and the 'near fatal' climax came for Springhill when the historic Bank Moratorium Directive came from the White House in March, 1933, closing every banking institution in the United States, which was followed in a few months by the suspension of operations by the Pine Woods Lumber Company after almost 40 years of continuous operation of their plant which was the backbone of the local financial structure. Hard times were indeed on all of us.[6]
>
> The Commercial Bank and Trust Company, after operating a year on the restricted basis, and failing in every effort to again open its doors, finally closed for liquidation on April 10, 1934. It was then that Springhill reached its lowest ebb. With its only industry gone, leaving hundreds without jobs, with the farm products glutting the market with no takers, with no banking facilities, Springhill was indeed struggling in the throes of its worst days. We had hope, and almost hope only left.[7]

Victory gardens, commodities shortages, insecurity, worry and fear, job losses, and business failures were ever-present in the nation and in Springhill.

Many of the residents of Springhill lived on farms or had a garden and a few livestock around the house. They told the author, "We did not have many material things, but we had food. We really did not suffer too much during the Depression." However, those who had no produce suffered like the ones Kate Butler described.

Springhill began to rebound when Frost Lumber Industries, Inc. acquired the holdings of Pine Woods Lumber Company in 1935. With another payroll after years of none, a gradual lifting of the clouds of the Depression came to Springhill. People sensed the end of a terribly dark period of history.[8]

In the summer of 1937 Springhill was selected as the site for a paper mill that would be owned by International Paper Company. Hundreds of workers came to town. Almost overnight Springhill was transformed into a boomtown. The Great Depression was over. The forest survived the stress.

War Years

The writer was a young lad during the war years 1939–1945. Our house was located on the present site of the Spring Theatre east of the L & A Railroad tracks. Frequently, trains loaded with military hardware passed through town. Tanks, military trucks, canon, troop transports and soldiers fascinated a young boy who often watched the passing trains from his back lawn. The town practiced blackouts in case there was a bombing raid someday. A small plane would circle the area for a few minutes, then the lights would be turned on. It was an eerie feeling for a small boy.

Years later, when I was a student at Baylor University studying history, I realized the horror of World War II. Normandy, Iwo Jima, Corregidor, Pearl Harbor, Tobruk and El Alamein became more than places on a map. They were killing fields where the Allies fought Hitler, Mussolini, and Tojo to preserve our democratic form of life. Hiroshima and Nagasaki became more than Japanese cities. Their names became associated with the new atomic era.

Global causes of World War I. The first of two great convulsions that tore Europe apart and much of the civilized world apart in the 20th century began on August 1, 1914. David Lloyd George, England's Chancellor of the Exchequer, said, "I felt like a man standing on a planet that had been suddenly wrenched from its orbit by a demonic hand and that was spinning wildly into the unknown."[9]

What began as a minor conflict when the Austrian Archduke was assassinated by a Serbian nationalist at Sarajevo in June 1914 had escalated into the "Great War." This was characterized as "The War to End All Wars." It was supposed to have been fought to save the world for democracy. After four years of battle, often in trenches, armistice came on November 11, 1918. Over 8.5 million lives were lost.

The war came to be regarded as a fusion of two different conflicts, each of which had a thousand years of history behind it: the conflict between the Teutonic and Slavic people for domination of Eastern Europe, and the conflict between Germany and the Franco-British alliance for the domination of Western Europe.[10]

The United States was brought into the war as a result of the sinking of three U.S. merchant ships, and by the Zimmerman Telegram. This was a secret message from the German foreign secretary proposing a German–Mexican alliance against the United States.

The end of the war ushered in the League of Nations, but the Allied powers negotiated a harsh surrender for Germany. One of Hitler's reasons for launching World War II was to avenge the armistice and Versailles Treaty of World War I. He began his war of revenge twenty-one years later.

Veterans of World War I. Springhill sent several of her sons into military service during the war: Eugene Basham, Tom Blount, George Coyle, L. A. Dunnigan, Graham Holland, Luther Holland, Andrew Bryan Holland, Gene Plunkett, and Jim Rowland are the names given to the writer.

Global causes of World War II. The Great War was begun when Hitler demanded and received the Sudetenland because of Neville Chamberlain's appeasement. The land had three million German inhabitants that Hitler wanted to integrate into the German nation.

Germany continued her aggressive behavior by invading Poland. This act caused England to declare war on Germany because of treaty obligations. Hitler used blitzkrieg or "lightning war" when he invaded Holland, Belgium, and France. Later he invaded Russia only to be defeated for many of the same reasons Napoleon was defeated when he invaded Russia.

During World War II 55 million men, women, and children died. Aerial bombardment flattened entire cities. Centuries of man's noblest accomplishments in art were consumed by fire from battles. Added to this was the horror of six million Jews, communists, and assorted liberals and intellectuals murdered in German prison camps. Man's inhumanity to man was practiced all over Europe.

Effects on Springhill. This great global conflict had a profound effect on Springhill. Gasoline was rationed, automobile tires were difficult to purchase, certain commodities were not available, and sons and daughters were sent off to war. Family members were separated from one another for a long period of time.

International Paper Company produced a certain grade of paper that was used for maps for General Eisenhower's army. Billy Baucum of Springhill used one of these maps when he was a soldier in the European theatre of war.

Veterans from Springhill. The most profound effect of the second world war on Springhill was the recruitment of citizens to go to war in Europe, North Africa, Italy, and the Pacific islands. Over two-hundred citizens from Springhill were in the military during World War II. Those citizens from Springhill who served the United States during World War II are listed in Appendix J.

Other persons who lived in Springhill after the war served in the military. They were drafted from their hometowns.

Four Veterans' Experiences

James Milton White—submarine casualty. James Milton White was a crewmember of the submarine "Tang." The crew had been firing torpedoes at the Japanese fleet. All torpedoes had been fired except two. They fired one of the two at a Japanese destroyer, then fired the last torpedo on the ship. Something in the guidance system went wrong. The lethal weapon turned back toward the submarine, blew a hole in it and caused it to sink. All crewmembers were killed. On board was James Milton White from Springhill, Louisiana.[11]

On Memorial Day, 2000, the American Legion honored James Milton White in a Memorial Service at the City Park in Springhill.

William T. Bowen—captured first Japanese prisoner on Iwo Jima. Toxie Bowen is the grandson of Dr. McDonald, the first medical doctor in Springhill. He had a distinguished career in the Marine Corps, serving in three wars: World War II, Korean Conflict, and Vietnam Action.

During World War II Toxie Bowen was a part of the invasion force that stormed the beach at Iwo Jima. During the bloody fighting in one of the toughest battles of the war Toxie captured the first Japanese prisoner on the island.[12]

Robert Charles Smith—National Commander American Legion. Robert Charles Smith was a native of Springhill, a 1935 graduate of Springhill High School, and the second highest official at International Paper Company. While he worked at International Paper Company, he was encouraged to pursue advancement in the American Legion. Smith had joined the Legion after James Allen took him to a fish fry in an effort to recruit him. His first post with the American Legion was that of athletic director of Banks-Strong Post No. 166.

He moved up the officer ladder rapidly while serving as finance officer, post commander for three years, vice-commander of the 4th Congressional District, district commander and state executive committee; state vice-commander, state commander of Louisiana, Louisiana's Alternate National executive committee man, International Affairs Commission for Louisiana for fifteen years and chairman of the commission for seven years; national vice-commander from the SSE area and national executive committee man.

Robert Charles Smith was elected National Commander of the American Legion in 1977 by unanimous vote. As National Commander he represented Louisiana as well as Springhill in all fifty states, Korea, Taiwan, Central America, France, Germany, and Flander's Field. He was proud to say, "I'm from Louisiana and I live in Springhill." He often met with the President of the United States and spoke over national television on behalf of all veterans, widows, and orphans.[13]

D. C. Wimberly—prisoner of war. D. C. Wimberly[14] was in the U.S. Army serving in the European theatre when he became a prisoner of war on Thanksgiving Day, November 25, 1944. He remained a prisoner of the German Army until May 15, 1945.

His capture occurred when his platoon was assigned to a twenty square mile wooded area for the purpose of killing or capturing artillery observers and snipers of the German Army. During the battle with the retreating army Wimberly entered a hot house building where flowers were grown. During this time the Germans shelled the area with "screaming meemies," 88 artillery firepower, and German tiger tanks. When the barrage stopped, he heard the sound of boots on the cobblestone. Another fight occurred. He crawled to the door, opened it and sprayed the German patrol with rifle fire. Then he jumped into a nearby woodpile to hide from the enemy. It was 1:00 A.M. By 2:30 A.M. the German patrol had captured three of his platoon members: Shelton, Smith, and Brown.

When the Germans began marching the three American captives away, Brown looked at the woodpile and said, "Come on Sarge." The Germans discovered Wimberly in the woodpile, hit him with a rifle on his back, and marched him several miles to a cave. He had been captured by Hitler Youth soldiers.

Treatment was harsh. They were marched for days without food, stoned and spit upon by civilians, and transported in overcrowded box cars across Germany without water or toilet facilities. Food on the train was two loaves of sawdust barley bread for all the men in the boxcar. Two cans of salted horsemeat were added to the diet.

He was taken to Stalag 3B in Furstenburg, Germany, on the Oder River. He needed help to get from the boxcar to the camp. During the journey from the woodpile where he was captured to Furstenburg where he was housed as prisoner, the Americans were marched day after day through heavy snow with little clothing. Six men would huddle together at night in the snow to attempt to keep from freezing to death. Wimberly said, "Food in Stalag 3B was made of grass, peas, or potatoes. A spoonful of sugar was provided once a week. There was one meal a day without salt that included 40 loaves of bread for 400 men. I ate charcoal for dysentery." [15]

Wimberly's release began when the Russians dropped artillery shells on the camp January 18, 1945. On April 22, 1945, Russians entered the camp at Lukenwald where he had been transferred by the Germans. He was liberated! The Russians carried them to Potsdam south of Berlin where he saw thousands of bodies stacked like pulpwood. The smell was sickening.[16]

He arrived home to meet his wife on June 29, 1945, and was discharged from the Army on December 12, 1945. When he returned to the United States he became involved with a new organization of ex-prisoners of war on September 8, 1963. In 1974-75 the group honored him by electing him National Commander of the American Ex-Prisoners of War organization. It now has a Congressional charter and offices in the Veterans Administration building in Washington.

The writer asked D. C. Wimberly a question, "How did you feel about the way the Germans treated you?" He responded, "When you lose your freedom, you become helpless. You are a slave. You survive by living by the second and thinking positive. You don't give up."

Killed in action. Nine Springhill citizens lost their lives in the war. They were: Robert Barnes, Burton B. Boyette, James H. Carmack, Jr., Charles W. Cason, and Buddy Chastant; Carlton Ferguson, S. J. Newbourne, Roy Wise, and James W. Whipple, Jr.

In Flander's Fields. Lieutenant Colonel John McCrae died in France while serving in the Canadian Medical Corp, and is buried in Winereaux Communal Cemetery. Before he died he wrote a poem about those who died in action during World War I. It has become perhaps the most famous poem about war. The poem is presented here as a memorial to all Springhill veterans who died while serving their country.

IN FLANDER'S FIELDS

In Flander's fields the poppies blow
Between the crosses, row on row,
That mark our place; and in the sky
The larks, still bravely singing, fly,
Scarce heard amid the guns below.
We are the dead. Short days ago
We lived, felt dawn, saw sunset glow,
Loved and were loved, and now we lie
In Flander's fields.

Take up your quarrel with the foe:
To you from falling hands we throw
The torch; be yours to hold it high!
If ye break faith with us who die.
We shall not sleep, though poppies grow
In Flander's fields.[17]

—John McCrae 1872–1918

William Luther Sevier Holland, U.S. Army, 1918–1919.

Andrew Bryan Holland, U.S. Army, 1918–1919.

Edward Graham Jackson Holland, U.S. Army, France, 1918–1919.

Robert Charles Smith, National Commander The American Legion, 1977–1978.

D. C. Wimberly, National Commander American Ex-Prisoners of War, 1974–1975.

Chapter 16

Diversity in the Forest–African-American Culture

The story of the African-American culture in Springhill cannot be separated from the African-American experience in America. Four eras identified the life of the African-American: slavery, emancipation, segregation and integration. In Springhill slavery and emancipation were not issues. These two experiences among African-Americans occurred before the city was founded in 1896. However, segregation and integration were woven into the fabric of the city from its beginning. The aftermath of events known in American history as the Civil War and Reconstruction left their mark on the attitudes of citizens regarding the African-American culture.

African-American Viewpoints of African-American Culture

Victimization. Women of the Black Caucus verbally protested the Electoral College votes in the joint session of Congress on January 6, 2001.

This was the continuation of the Martin Luther King non-violent protest tactic used across the nation since the mid–1950s. Being against an issue and protesting has been the political action strategy of the African-Americans. Mr. Richard Marshall, the writer, stated that African-Americans felt victimized because of slavery and discrimination which led them to seek redress through affirmative action, welfare programs, educational grants, reparation, and Justice Department law suits.[1]

Dr. John McWhorter, an African-American professor at the University of Southern California at Berkley expanded on this popular African-American viewpoint about themselves and about their culture. He wrote in *Losing the Race* that many African-Americans felt and believed their lives resulted in the cult of victimization, separation from mainstream America, and anti-intellectualism. He wrote, "There is joy in underdogism."[2] Jesse Jackson promotes this concept with the tactics of non-violent protest, affirmative action, and law suits in his efforts to redress wrongs.

Dr. John McWhorter believes the era of protest is over for his people. Civil rights laws and voting rights laws have been passed. He believes it is time to take responsibility for education, family life, and business opportunity. He said, "It is time to move on. Many African-Americans talk about personal responsibility, education, family life, and working hard for accomplishments in private conversations, but it has not been accepted by African-Americans in the public forum."[3] He promotes the thesis that African-Americans are losing the race in America. It is expressing itself in three manifestations:

The first is the Cult of Victimology, under which it becomes a keystone of cultural blackness to treat victimhood not as a problem to be solved, but as an identity to be nurtured. The second manifestation is Separatism, a natural outgrowth of Victimology, which encourages Black Americans to conceive of Black people as an unofficial sover-eign entity, within which the rules of other Americans are expected to follow are suspended out of a belief that our victimhood renders us morally exempt from them. Separatism spawns the third manifestation, a strong tendency toward Anti-Intellectulism at all levels of the Black community. Founded in the roots of poverty and disenfranchisement, this tendency has now become a culture-internal infection nurtured by a distrust of the former oppressor."[4]

What will make things better? Public demonstrations are not the answer from his perspective. He thinks Anglo-Americans need to continue to work on their prejudices and attitudes of racism, but African-Americans need to get beyond the feeling of "victimization." Education is the heart of the answer to win the race. Dr. McWhorter believes many African-Americans now accept the view that they are responsible for making things better for the African-American culture.

Springhill has been influenced by both views. Public demonstrations have not been held, but the victimization concept is felt among many African-Americans. On the other side of the debate are African-Americans who work hard, become involved in American capitalism, and become educated. They have been influenced by the African-American culture, but the contemporary thinking among many encourages personal responsibility and education.

Historical perspective. A review of the eras of African-American culture will provide a perspective of African-American life in Springhill.

Slavery. African-American plantation slavery began in the New World in 1517 when Spaniards began importing slaves from Africa to replace Indians who died from harsh working conditions and exposure to disease. In 1619 a Dutch ship with 20 Africans aboard arrived in the colony of Jamestown, Virginia. Five hundred slaves arrived in Louisiana in 1719 to clear the fields, plant and harvest the profitable crops of indigo and cotton, and take care of the plantation.[5]

Slaves were considered property rather than persons. They had no freedom, were bought and sold for economic purposes, and were totally controlled by their owners.

Emancipation. The Civil War in the United States began in 1861 in Charleston, South Carolina, as the Confederates opened fire on Fort Sumter. It ended April 26, 1865, after the surrender of Confederate Generals Robert E. Lee and J. E. Johnson. Two issues were basic causes of the war: economic conflict over slavery between the

North and the South and the South's feeling that it was becoming a minority and losing political control.[6]

President Lincoln signed the Emancipation Proclamation on January 1, 1863, which gave freedom to the slaves. Reconstruction in the South followed the war until the last federal troops were withdrawn in 1877.

Segregation. Bitterness lingered long after the Civil War ended. In the South separate water fountains, schools, eating places, recreation facilities, and hotels were required by the white citizens. At one time the State Theatre on Main Street in Springhill had a small segregated place in the balcony for African-Americans. The public schools in the city were like all others in the South—segregated. African-Americans would come in a white person's house through the back door if invited. African-Americans were not allowed to vote until the Voting Rights Act was passed in 1965 following the Selma to Montgomery march. These things occurred in Springhill.

In order to correct these social and political problems groups organized to redress grievances. A group of whites had formed the National Association for the Advancement of Colored People (NAACP) when W.E.B. DuBois' Niagara Movement joined with them in 1908.[7]

In 1942 the Congress of Racial Equality was founded in New York City. (CORE). In 1957 the Southern Christian Leadership Conference was established by Martin Luther King, Jr. to assist local organizations working for the full equality of African-Americans.

These events and organizations influenced segregation in Springhill. Split Log Quarters and Saw Mill Quarters were the designated housing areas for the African-Americans. They were located near the present-day Brown Junior High School. The national stress over racial issues created similar viewpoints in Springhill but only in recent years has there been an organized NAACP.

Integration. A major change occurred in racial issues in 1964. President Lyndon Johnson signed the Civil Rights Act, giving federal law enforcement agencies the power to prevent racial discrimination in employment, voting, and the use of public facilities.[8] On May 17, 1954, the Supreme Court ruled unanimously in Brown v. Board of Education of Topeka that racial segregation in public schools violates the Fourteenth Amendment to the Constitution. In 1970 the Webster Parish School Board integrated schools in the parish. Brown High School, a segregated Black school, was integrated. In Springhill the integration of schools was accomplished by a combined zoning and a consolidated plan. Teachers and students in all schools were integrated.

Summary. Attitudes and actions regarding segregation and integration in Springhill followed the national pattern. Prejudice and racism have been, and remain, part of the psyche of the city. However, major progress has been made in racial issues. Integration of schools has been accepted, the Civil Rights Law of the land is followed, African-Americans are involved in the City Council and Police Jury. African-American businessmen are successful entrepreneurs, and African-Americans have moved out of the "quarters" into mixed neighborhoods. Although there are strong disagreements about integration, citizens have accepted it as the law and are attempting to live in racial harmony.

Sociological Conditions

Housing has improved dramatically for African-Americans in Springhill. Although "shotgun" houses remain, many in run-down condition, there are many African-Americans living in middle class neighborhoods. Camelot, a Section 8 housing complex south of Springhill High School, is an attractive and well-planned area where many African-Americans live.

The paper mill influenced the pay scale for African-Americans when it was constructed in 1937. Wages increased tremendously. Since that time economic conditions have improved for African-Americans in the city. Lawn care, printing and graphics, construction, and institutional service jobs are examples of businesses in which African-Americans are involved.

Political conditions have opened for African-American participation. Ed Bankhead is on the City Council, Jimmy Thomas is on the Police Jury, Jonathan Washington is President of the local NAACP and Malachi Ridgel is on the Webster Parish School Board. Henry Rhone is a wealthy African-American businessman.

Religion is important and influential in the African-American community. Harrison Chapel, Washington Church, and New Bethel AME are strong advocates of ministry and community participation. Rev. Jonathan Washington organized and developed the Springhill Community Activities Center near Brown school. It is used for meetings of various kinds, social activities, and church-related events.

Future African-American Culture

African-Americans in Springhill have moved with the American African-American experience from segregation to integration to economic, political and educational involvement. African-Americans are now experiencing the American dream even in the midst of prejudice. They, along with others, are in the growing minorities of the nation. It is probable that their influence and involvement will continue to increase locally as national influence and involvement grows.

First school building.

Second school building.

Left: *Charles Herbert Brown (1914–1951), for whom schools were named.*

Above: *Charles H. Brown Middle School.*

Below: *Entrance to Charles H. Brown Middle School.*

Charles H. Brown Middle School, 2001.

North-South Construction Company (former Sanitary Dairy Building.

Alma Mater

Tune: Auld Lang Syne

Dear Charles Brown High, we'll ne'er forget;
We pledge our love to thee
The green and white we'll ne'er neglect,
Through all eternity.

CHORUS

To honor Charles Brown High Schoolmates
To honor Charles Brown High
We'll always be devoted to
Our dear ole Charles Brown High.

Through all the years, we loved thee best
Our faith in thee remains
To you we owe our happiness,
We'll return to you again.

Alma Mater page from "The Tiger," a history of the Brown schools.

Chapter 17

The Neighboring Forest– Cullen, Sarepta, and Shongaloo

The three Northwest Louisiana towns of Cullen, Sarepta, and Shongaloo have an identity of their own. They also have a rich heritage. Nevertheless, they are neighbors with Springhill, interlocked with some common characteristics. The L & A Railroad, timber industry, and International Paper Company are shared interests by Cullen, Sarepta, and Springhill. Timber and oil are common interests to Shongaloo and Springhill. Related families live in the four towns. They are neighbors in the forests.

Cullen—Pulpwood Town

Early settlers. Hugh J. Coyle and his wife Louvenia Braley Coyle built their home in 1878 on what is now known as Pecan Street. Mr. Coyle was a cotton farmer involved in buying, selling, and financing cotton crops for local farmers. He also owned and operated a general merchandise store.[1]

Coyle store. When the L & A Railroad was extended from Taylor, Arkansas, to Alexandria, Louisiana, via Springhill and Clifford (Cullen), a flag stop, Mr. Coyle moved his store near the railroad. It became a shipping center for cotton, a depot, and a United States Post Office. Cotton from the buying lot was transported to Shreveport for sale by horse and wagon. Stock for the store was loaded into the wagon for the return trip to Clifford. The store and post office closed in Clifford in 1914 and was moved to Springhill.

Mail service. The Coyle store had been the post office that received mail by pony express. When the store was moved to Springhill, the post office was closed in Clifford. The community received mail from Sarepta until the late 1930s.[2]

School. When the Coyle store closed and moved, children were attending elementary school at King's Corner or Springhill. Each family provided for their children's transportation or else they boarded them with friends or relatives near the school.

International Paper Company. In 1937 International Paper Company purchased ninety acres of land from the Coyle estate. Overnight Clifford was flooded with construction workers and their families. Temporary tents and small dwellings sprang up in every available space.

A depot was constructed to handle incoming and outgoing freight for the mill. The paper mill began manufacturing paper in 1938. Clifford continued to grow as the number of employees at the mill increased.

Businesses. Mr. and Mrs. Jennings Cassells opened a small grocery store in their residence on Coyle Street during this time. Jack and Mac Curtis expanded their grocery store into Clifford in 1937 and located it where the present Be Bop Grocery operates. Their business grew to the point that they could expand to a second store in Clifford. They purchased it from Tim Ford. Jack operated the second store and Mac operated the Highway 7 store.

Churches. As the community of Clifford grew, people felt the need for spiritual fulfillment. Thus, churches began to be organized and attended. The Church of God was organized in 1938 with the Rev. Floyd Webb as pastor. First Baptist Church was organized in January 1939 with Elder Errol Williams as pastor. In 1945 Calvary Baptist Church was organized with the Rev. Carol Peaden as pastor. The Colored First Baptist Church was organized in 1938 with Rev. Woodson as pastor.

Other churches located in Clifford include Evening Star Methodist Church, Church of God in Christ, and Holy Temple Church of God in Christ.[3]

Name change. As a result of International Paper Company locating in the area and in honor of the president of the company, the name Clifford was officially changed to Cullen in 1938.[4]

Richard J. Cullen possibly did more than any other man to further the development of the paper industry in the South. He was one of the first to open the vast resources of the Southern woodlands for the raw material to make kraft paper. After a successful career as a mechanical engineer building paper mills in Bastrop, Louisiana, Camden, Arkansas, Moss Point, Mississippi, and Mobile, Alabama, he became President of International Paper Company in 1936.

Incorporation. By proclamation of Governor Robert Kennon, Cullen became an incorporated town June 18, 1955, making it the youngest incorporated town in Webster Parish. The approximate population at that time was 1,600 people.

The Governor appointed the following officials: M.C. Burnham, Mayor; M. A. Curtis, A. A. Halterman, Fred Dreher, and J. D. Curtis, Aldermen; J. W. Brown became town marshall.

The first meeting of Mayor Burnham and the Board of Aldermen was August 1, 1955. Town hall offices were set up in the office of Curtis Brothers store. Later they moved the offices to the Burnham building. A. A. Halterman acted as clerk until November 1955 when Bernice Bradshaw was employed as town clerk and tax collector.

The first election was held in the spring of 1956. Winners taking office on July 1, 1956, were M. C. Burnham, mayor; aldermen J. D. Curtis, Jasper Burt, A. A. Halterman, S. L. Lout, and Lloyd Stewart; Leon Pafford was Town Marshall.[5]

Civic clubs. From time to time civic clubs were organized, each contributing to the development of Cullen. The Cullen Lions Club was chartered in March, 1960. Ed

Morris served as the club's first president. The businessmen of Cullen joined with the Springhill Chamber of Commerce and became known as the Cullen-Springhill Chamber of Commerce.

Present-day Cullen. Sandra Brown, staff writer for the Minden Press-Herald, wrote, "Even though it is basically connected to Springhill and International Paper Company, Cullen takes care of itself and has its own identity." [6]

When International Paper Company closed the mill in 1979, Cullen suffered. Although its identity dates back to 1878, present-day Cullen is closely tied to International Paper Company that is located across the highway from the downtown area. Many adjustments have been made.

During the eight years following incorporation in 1955, the town made great strides. It built a water distribution center, erected a 100,000 gallon storage tank, drilled two deep water wells, installed a sewage collection system, and completed an oxidation pond for sewage disposal. In the midst of these expenditures Cullen maintained a favorable financial position. Over seventy-five per cent of the streets were black topped and maintained. Fire and police protection were excellent.[7]

Today a new full service truck stop owned by Kenny Coyle and Kenny Coyle, Jr. has opened. Be Bops Grocery and Market, Rhones Dry Goods, Sams Mens Wear, and Regions Bank are in operation.[8]

Chief Dexter Turner has turned the police department around and the crime rate has dropped. There is a computer system, two offices, four policemen, and four patrol cars. Turner said, "After living in Cullen for two years, I realized there needed to be changes. I saw the department as not having direction—and I thought I could give it some leadership. Before, we were infiltrated with drug dealers and the crime rate was high, but with our patrolling and the help of the sheriff's office, we have got a hold on it." [9]

Bobby Washington was Mayor of Cullen for twelve years. His last day in office was December 31, 2000. In these twelve years he saw many changes. He believes one of these changes is most important. Mayor Washington said, "Hopefully the attitude of the people has changed to realize we are going to have to get up and do things instead of depending on somebody to give us something." [10] This is the new Cullen. The town's current mayor is Floydean White.

Sarepta—Religion and Railroads

Jennifer Vargin, staffwriter for the *Minden Press–Herald,* visited and wrote about Sarepta for the paper. On her drive back to Minden Ms. Vargin attempted to sort out her feelings of why she disliked small towns. "One of the reasons I don't like small towns is because I do not feel they are adequately equipped to handle life's emergencies. And there are not enough places to visit or shop, and entertainment is hard to come by." [11] Then she wrote, "But I realized something profound. People like it here because they do feel secure. They have each other to depend on and don't worry about finding help in emergencies because they always know someone will be there for them." [12]

"Sarepta is a place where children are safe, crime is extremely low, and community support is outstanding. People are happy there because even if they work in other cities or towns, they can come home to the peace of Sarepta and find sanctuary."[13] In Sarepta school is important, family is important, and religious faith is important. This is their history.

Early settlers. According to the Webster Parish land records, Township 22, Range 10 West, Section 7, in which Sarepta is now located, was first settled on October 27, 1848, by William Denham who purchased one-hundred sixty acres. Two years later he acquired neighbors when James Butler and Simon D. Moore settled near the Denhams.[14]

Schools. Five years later in 1853, the first school in the community was established a few hundred feet from Mount Moriah Church on Highway 7 where the Church of Christ now stands. It was a one-room log cabin heated by a large open fireplace. The first teacher was C. C. Martin.[15]

Several years later the school moved to a location about 300 yards from the home of L. L. Houston. The school burned but new school construction was immediately launched. In April 1918 the school board voted to consolidate Sarepta and Clifford schools. A new building was erected for the new high school. The class of 1923 was the first one to graduate from Sarepta High School. This Sarepta school, with all its equipment, was totally destroyed by fire October 27, 1927. A new brick building was constructed. In the fall of 1928 students had a new school building in which to attend classes.

Name. Sarepta is a biblical name found in *Luke* 4:26 (KJV). It was a Phoenician town to which Elijah was sent during the time of a great famine in order to save the lives of a widow and her son (*I Kings* 17:9-10). The Revised Standard Version of the *Bible* adopted the form of the name based upon the name in the Hebrew scriptures, "Zarephath." Josephus the Jewish historian, places it between Tyre and Sidon along the Mediterranean seacoast. Christ probably visited the city (*Mark* 7:24–31).[16]

The name of the town "Sarepta" came from Mrs. Sarepta Carter. When the congregation was deciding on a name for the church that Mrs. Carter attended, she offered to buy a large pulpit *Bible* if the members would name the church in her honor. "Being a Christian woman of the noblest type," the church members considered it an honor to use her given name, "Sarepta," for the name of the church. The year was 1868. The offer of Mrs. Carter was gladly accepted. The Pulpit Bible, the gift from Mrs. Carter, was installed in its proper place and the rural church became known far and wide as Sarepta Church.[17]

Ten years after the construction of the new building for Sarepta Church, Mrs. A. J. Allen built a small general store one-half mile from the church. The spot is known to older citizens as "The Old Hickory Tree," a common meeting place for the citizens. The store soon became known as "the store at old Sarepta." The small settlement which now consisted of the church, a cemetery, a store, and a few small farmhouses was called Sarepta. Mrs. Carter's dream of being remembered became secure as the little village grew to the present-day Sarepta.

The name "Sarepta" was formalized in 1884 when a post office was established by the United States government. It became known as Sarepta.

L & A Railroad. William Buchanan, who had built Bodcaw Lumber Company in

Springhill in 1896, owned and operated the L & A Railroad from Stamps, Arkansas, to Vidalia, Louisiana. He used the railroad to transport his lumber to market.

In 1898, twelve years after the construction of Mr. Allen's store at the "Old Hickory Tree," the L & A Railroad succeeded in securing the right-of-way from Mr. J. I. Allen and started work on a railroad that would pass 150 yards from the store. Sarepta then began to grow.

Stores were constructed by J. S. Sikes in 1902, M. L. Cox, J. A. Robinson, Coyle and Company in 1905, Dr. C. T. DeLoach in 1918, and R. B. Kennedy in 1918.

In 1920 the Bank of Sarepta was organized. In 1922–1923 business establishments owned by George Tyler, Robert Steel, Grover C. Allen, Standard Oil Wholesale, and the Telephone Exchange were constructed.

Incorporation. The town was incorporated in 1955. Richard Griffin was elected the first mayor.

Trace Atkins. Sarepta's most famous son is Trace Atkins. He was known as a good hard-working student, and a person with an interest in music. For his third Christmas his parents gave him a toy guitar. When he was ten years old, he received a real guitar and began guitar lessons.

In high school Atkins became a member of a gospel quartet. Danny Reeves was a member of the quartet and encouraged him to sing. He moved to Nashville in 1991 and became a construction worker. Scott Hendricks, CEO of Capital Records heard Atkins sing at Tilley's and immediately offered him a contract. Shortly after that event, Atkins produced his first album, "Dreaming Out Loud," from which he enjoyed three number one hits. Later he produced another album, "Big Time." In 1999 he landed a contract with Chevrolet.

He has said his childhood in Sarepta influenced his music. Sarepta's home town boy who has become very successful in the country music world continues to be a source of pride for his strongest fans—his Sarepta friends.[18]

Leadership. The high school is led by principal Steve Fowler. Howard Beaty, Jr. is vice-president and branch manager of Regions Bank in Sarepta. Pam Dorsey is mayor. Steve Wooley is a pastor who gives spiritual leadership to the citizens. People of all ages like Sarepta. When asked about their town, they say, "I'm happy with Sarepta. I can't think of anything I would change."

The Old Oak Tree. This tree is located one-half mile north of Sarepta on the property of Mr. T. Caraway. It is on the road from Sarepta to Shongaloo. The tree marks the place where the first village store was built by J. L. Allen.

The old hickory tree became famous because it was designated as the meeting place of the neighborhood and of hunters and campers on their way to their outdoor activities. Individuals would say, "I will meet you at the Old Hickory Tree." The tree became a symbol of neighbors and friendliness in the village and remains today as a part of the folklore of Sarepta.

Shongaloo—Cypress Trees and Running Water

Charming, quaint, serene, tranquil, beautiful—all are words that citizens use to describe the little village located at Louisiana Highways 2 and 159 half-way between

Magnolia, Arkansas, and Minden, Louisiana. Shongaloo is one of the oldest settlements in Webster Parish.

Name. The name Shongaloo is an Indian name—"shakain"—which means "cypress trees" or "running water." Old Shongaloo was a Caddo Indian village called "Shakalo." The Caddoans lived in Southwest Arkansas, Northwest Louisiana, and Northeast Texas along the Ouachita River and Red River from 800 to 1835 A.D. They were advanced in their culture, often living near streams or rivers, farming the land, hunting and making artifacts. Their villages were united into several confederations and ranged the area near Hot Springs, Arkansas, to Natchitoches, Louisiana. They had a strong government for the time in which they lived. (See Chapter 2). When white settlers came to the area the name was changed to Shongaloo and later to Old and New Shongaloo.

Early settlers. White settlers began moving into the area in 1814, although slowly, because it was a tangled mass of vines and thick woods. There were no roads—only Indian trails—and no way to get wagons through the underbrush and woods. Eventually, the great fires of 1824 cleared much of the land for settlement.

In 1803 President Thomas Jefferson bought Louisiana from France for $15 million. At this time migration from Tennessee and from the southeastern part of the United States ensued. In 1814 John Turner Sikes, born in Liverpool, England, came to Northwest Louisiana, where he built a trading post and a ferry across Dauchite (Dorcheat) creek at an old Indian crossing.[19] In 1856 the public rates for the use of the ferry were set at: 1 wagon and 4–6 horses or oxen—$1.00; 2 horses and carriage—.75; 1 horse wagon and carriage–.50; man and horse—.25; footman—.05; cattle per head—.05; sheep and dogs per head—.03. The ferry, which was built in 1816, operated until 1910 when the first bridge was built. At this time William Darby's 1816 map shows only a wilderness with Indian trails and only one settlement—a Caddo village called Shakalo.

During the next few decades settlers came slowly but steadily in two migration patterns—one from Middle Tennessee, down the Mississippi River, up the Red River into the forests of Northwest Louisiana; the other came from the southeastern section of the United States. In 1820 Boots was settled and later became known as Minden. In 1828 Claiborne Parish was founded. It included the present-day parishes of Bienville, Union, Lincoln, Webster and Claiborne. In 1850 a store and sawmill were begun east of the village of Shakalo.

In the mid 1850s several sons of James and Permelia Wise came to Shongaloo from Gordon, Louisiana, and settled near the Indian village. The place became known as Wiseville which was the site of the first post office that opened August 8, 1857.[20] Dr. Giles James Wise practiced medicine at the Indian village and in Wiseville.

Old and New Shongaloo. Doyles Crossroads, named after Major John Doyles, was homesteaded by William D. Harrison in 1880. He nailed a sign over the door of the store located in the Northeast corner of the crossing and called it New Shongaloo. On August 10, 1881, the New Shongaloo post office was established.

As white settlers moved in, Caddo Indians moved out, and the settlements became known as Old Shongaloo and New Shongaloo.

Incorporation. Shongaloo became an incorporated village in 1967. John Brady was appointed mayor. J. C. Bridwell, Stanley Farmer, and Bachmon Atkins were appointed aldermen. Parey Branton, state representative, delivered their commission from Governor McKeithen.[21]

Economy. Four keys unlock the economic stability of Shongaloo: oil and gas exploration, agriculture, timber, and small businesses. A few citizens commute to jobs in other cities.

In 1921 the Louisiana Oil Refining Company discovered oil on the W. E. Gleason property. Other fields were discovered as follows: Haynesville, 1921; Homer, 1921; Cotton Valley, 1922; and Springhill, 1951. The oil industry has contributed greatly to the wealth of the community.

Agriculture is also an important economic factor for Shongaloo. Newsom Dairies is a family owned business. There are several chicken and egg farms. Cattle are raised in the area. The youth 4-H program has been an active one because of the agricultural emphasis in the town.

Other families do not consider themselves directly dependent on agriculture for their livelihood, but they farm small areas. Backyard gardens are an important part of the life of the town.

Timber is important to many landowners as an investment. Growing and harvesting pine trees is an economic development program for which the area is well suited because of the climatic factors—favorable temperature and adequate rainfall and growing season. The forests are important to hunters since they provide a natural habitat for numerous species of wildlife.

Small businesses have stocked merchandise in their stores to meet every need. These businesses have included Pixie Family Store, Clark Burnes store, Gladney Haynes store, Rice Copeland's store, Clark Burnes Downtown store; Loy Cox's store, Jimmy Lindsey barbershop, Robert Max Hayes barbershop, and Flo's Café; Vines Insurance Agency, Mac Dunn's Café, Roseberry's store, Shongaloo Grocery, and Indian Run Restaurant.

Education. The first school in the general area was built of logs in 1850 and was known as Union Springs School. Children's families paid tuition in the form of farm produce for two to three months of school.

Other schools were established in the following years: Matthews school, McEachern school, Sikes Ferry school, Hall school, Slack school, Potters school, and Pioneer school.[22]

In 1921 schools in the northern part of the area were consolidated into Old Shongaloo School and those in the southern area were consolidated to form the New Shongaloo School. The schools were later combined to form Shongaloo High School in the same year and gained state approval in 1922. The frame building was destroyed by fire in 1925. A new brick building was constructed in 1927. The present school buildings were completed in 1993. The school district encompasses the area from Dorcheat creek to Claiborne Parish and from the Arkansas state line midway to Minden.

Leadership. Many people have contributed to the growth and development of Shongaloo throughout its history. Included among these leaders are: Dr. G. J. Wise, Major John Doyles, William Harrison, Jesse Sikes, and William B. Roseberry; Parey Branton, Georgia Branton, and Mertis Young who served with distinction as mayor 1997–2000.

Parey Branton is perhaps Shongaloo's most distinguished son. He served as state representative 1960–1972, mayor 1972–1988, and Webster Parish school board president, 1960–1972.

Tom Craig, who attended Shongaloo schools and graduated from LSU, served with distinction in the United States Air Force. Throughout his career he had many important responsibilities with the Air Force and military planning. He retired with the rank of major-general.

Churches. The spiritual life of Shongaloo has been important throughout her history. The first church was built in 1845 by the Methodists. Craig Hearn was one of the earliest ministers to preach from the pulpit of the log structure. The present Methodist Church is located on the same site.

Pilgrim's Rest Church was established four miles east of Shongaloo in 1842 by Rev. Peter McDonald, a Methodist circuit rider. Both Baptists and Methodists worshipped there until the Methodists established Shady Grove Methodist Church in 1853.

In 1890 twenty-seven members of Pilgrim's Rest Church established a new congregation which they named Mount Paran Baptist Church. A new wooden structure that seated three-hundred people was built in 1926. The church was located on donated land up the east hill from the Shongaloo crossing. In 1854 a brick building was constructed by the congregation. Today, the beautiful chimes call villagers to worship on Sunday mornings.[23]

Other churches that have been established for the spiritual development of the people of the Shongaloo area are: Old Shongaloo Missionary Baptist Church, Old Shongaloo United Methodist Church, Old Union Baptist Church, Sweet Home Baptist Church, and Union Springs Baptist Church.

Interesting sites. A tourist to Shongaloo can experience the history of the village by visiting the Gleason Crater, Log Cabin Museum, Sikes Ferry, Indian Springs, and the Civic Center.

Gleason Crater was formed in 1921 when the drill hole for natural gas and oil "went wild." The slush pit fell to a depth of one hundred feet, the well fell into the pit and the derrick fell to the side. Fortunately, no one was injured when the force of nature erupted. The drill hole caught fire a year later and burned for two and one-half years. When Bub Martin's well was drilled nearby, pressure on the crater was relieved and the fire went out. Debris has filled in the crater to some degree, but a large hole created by a blown-out well remains.[24]

The Log Cabin Museum was constructed behind the Civic Center in 1976 for the Shongaloo Bicentennial Celebration. Three log cabins were dismantled and one single cabin was built from the logs. Parey Branton, then mayor, numbered and counted the logs as they were transferred. Betty Brown and Georgia Branton served as co-chairmen of the project for the Garden Club. Many persons in the community assisted in the construction and development of the museum. These included: Lonnie Martin, Clinton Martin, Darnell McEachern, C. C. Sexton, and Robert Slack; Jake and Verna Bridwell, Stanley Formby, Roger Hemphill, Roger Wise, and Rodney Wise; Elton Wise, Gussie and Drew Haynes, Mildred Burns, Mrs. Noble Leonard, and Mrs. Ever Martin; Mrs. Era Nunn, Mrs. Mary Will Matthews, Allie Mae Slack, and Jesse Mae Adkins.

The museum is filled with household furnishings that demonstrate how people lived in the backwoods of America in the 1800s.[25]

Conclusion. A reporter for the Minden Press-Herald wrote a story about Shongaloo for "Profile 2000" in their paper. She asked, "What is the first thing that comes to mind when you think of Shongaloo?" Many people responded with the word "home." She wrote, "I think to these folks 'home' means more than a place to live. It seems to me, from the youngest to the oldest, in Shongaloo it means commitment and pride."[26]

King's Corner Assembly of God Church.

The "Southern Belle" visits Cullen in March of 1949.

Main Street in Cullen about 1949.

Old Sarepta Missionary Baptist.

Sarepta Municipal Building.

Shongaloo United Methodist Church (the Rock Church).

Shongaloo School, built in 1993.

Sarepta High School.

New Sarepta Baptist Church.

Mt. Paran Baptist Church of Shongaloo.

First Baptist Church of Cullen, 2001.

The SPARC Center in Sarepta replaced the burned building on Spring Branch Rd. in Springhill.

Cullen Fire Station.

Shongaloo Civic Center and Veterans Memorial.

PART SIX

Pine Cones on the Tree—Folklore and Biography

Chapter 18

Folklore in Springhill

Folklore is defined as the traditional beliefs, myths, tales, and practices of a people transmitted orally.[1] A story is defined as an account or recital of an event or series of events, either true or fictitious.[2]

The following stories are characterized as folklore. They are written for the reader's enjoyment and for a historical record. There is no reason to group them or place them in any order. Read them consecutively or skip around through the stories. Some are included in the text of the book. Others come from citizens who shared them with the author. All of them tell us something about the history of Springhill.

Play from your heart. Ed Shultz loved the Lumberjack football team. When the 1952 "Jacks" played Reserve High School for the state championship, citizens did various things to show their support. Mr. Shultz had cards printed which read, "If you play from your heart, you can win." He passed them out to the team immediately before they ran on to the field to play for the state championship. It worked. Springhill beat Reserve 20 to 13 to become state champions.[3]

Piano playing at the silent movie. J. A. Branch and J. B. Branch owned the first silent picture show in town. Mrs. Branch popped popcorn in a pan for the customers. The film was silent, so Mrs. Branch or Jack Branch played the piano while the movie was being shown. This was part of the entertainment. The piano was located in the front of the movie house near the screen.[4]

Rats in Webster theatre. The Webster Theatre was located near the present site of the Lucky Dog Café on Main Street. This location was next door to a bar. Rats from the bar invaded the theatre after it closed for the day in order to eat the popcorn that had spilled on the floor. Mr. L. E. Berry, manager, went to the theatre early on Saturday to chase the rats away from the scattered popcorn, clean the seating area and prepare for the Saturday 10:00 A.M. movie.[5]

Oak slats to Europe. Frost Lumber Company manufactured oak slats that were used to build wooden wine vats in Spain and Italy. Local residents began stealing the oak slats for their personal usage. Frost Lumber Company solved the problem. They shipped the entire oak trees to Spain and Italy. Slat stealing stopped.[6]

Uncle George's Café. Uncle and Aunt George Essig owned a hamburger café at the corner of Giles Street and Church Street. They became famous for their hamburgers that Mrs. Essig mixed in a large pan. She usually added oats to the meat. Children, youth, and adults would go to Uncle George's Café in groups to eat ten cent hamburgers and drink five cent big oranges. Children and youth would visit the place during the noon hour lunch break. Adults attended the midnight show at the State theatre and then purchased hamburgers after the movie was over. At midnight there would be a line of adults walking up the hill toward the Arkansas highway eating the famous hamburgers.[7]

Board Minutes of Pine Woods Lumber Company. One gets the impression that board meetings of the lumber company were held and motions deliberated with all due consideration. Close scrutiny of the minutes reveals that only one man was in the room—J. F. Giles. He wrote the minutes from proxies and orders from the Texarkana office as if all the board members were present. He acted out the deliberations of an invisible board of directors. He must have become bored with the process because from 1901–1906 there were only blank pages in the minute book.[8]

Branch Brothers Motor Company. J. A. Branch came to Springhill in 1917 from Hornbeck, Louisiana, bringing his wife and young son, Jack. Mr. Branch became a meat cutter in the newly opened commissary of Pine Woods Lumber Company. He became a friend of J. F. Giles, the manager, whom he assisted in driving to Shreveport. When Mr. Giles purchased the Ford agency in 1922, he asked J. A. Branch to operate his motor department. Mr. Branch agreed on the condition that his brother J. B. Branch could help him. Mr. Giles agreed, so J. B. Branch and his brother-in-law, Ralph Lacey, and his wife came from Henderson, Texas, to investigate the situation. J. B. joined J. A. as parts manager and assistant manager for the Ford agency. The agency was located at the site of the now-closed Regions Drive Through Bank on the corner of Main Street and Highway 157. They moved to the present site on South Main Street to accommodate a growing business.[9]

Commissary script. William Buchanan owned the town. He would pay his workers in script that was redeemable at the commissary. However, company employees were tired of paying high prices at the commissary which had control of the market. One merchant on "The Hill" began accepting script from Buchanan's employees. This gave the commissary competition. When the merchant on "The Hill" submitted the script to Pine Woods Company in exchange for legal tender, the company had to comply. Pine Woods Commissary began reducing prices to meet the competition.[10]

Breakfast for loggers. Cooks would be in the kitchen by 2:00 A.M. preparing breakfast for the loggers who worked in the woods. A menu of biscuits, eggs, fried dry salt bacon, syrup, butter, and coffee was prepared for the men. The same menu was prepared and placed in a syrup bucket for each lumberjack to carry to the woods for lunch. A large bell by the kitchen would ring at 4:00 A.M. to call the men to breakfast. Trains would leave at 5:00 A.M. for the woods and return after 5:00 P.M. The men had to be at their work sites by 6:00 a.m. when the whistle blew.[11]

Mr. Giles' conservative ways. On a cold day Mr. Giles joined a group around the heater in the commissary. He said, "I want to give you this advice. When I was a bookkeeper at the Stamps mill, I smoked cigars—15 cents a day. When I married, my wife and I talked about our plans for the future and what we wanted to accomplish. We agreed that if I quit cigars and put the money aside, we would be able to buy a carpet next year. Every year we could think of something we needed more than I needed a cigar. Now we have a nice nest egg.[12]

Redeeming Coke bottles. Mr. J. F. Giles rode around town in a horse-drawn surrey. When he saw a Coke bottle on the side of the road, he stopped the horse, retrieved the Coke bottle and carried it back to the store to redeem it for cash. Mr. Giles died a millionaire. His fortune today would probably be $40,000,000.[13]

Chocolate ice cream on a white suit. When James Branch was a little boy, his mother dressed him in a white shirt and pants and carried him to the Pine Woods Commissary. The author's mother, who worked in the store, took care of him while his mother shopped. He was given a delicious chocolate ice cream cone which dripped all over his white clothes. His greatest fear was that his mother would not allow him to return to the commissary with her. However, she cleaned him up, carried him home, and returned with him on another day.[14]

Grant for North Webster Industrial District. There was a bureaucratic problem between agencies in Austin, Texas, and Washington regarding the approval of a grant for the planned North Webster Industrial District. Senator Roemer, who later became Governor of Louisiana, met with Charles McConnell and James Branch to discuss the status of the grant. Senator Roemer paid the agency in Austin a personal visit to encourage the executive of the agency to approve the grant immediately. This was accomplished. The grant money was released immediately and the North Webster Industrial District was established by using the funds from this grant and other funds that were available.[15]

Branch Brothers' telephone. The only telephone in Springhill that could be used privately was located in the office of Branch Brothers Motor Company. Business men would come to town, visit Branch Brothers, close the office door for privacy, and use the telephone for a business call. They would then leave the agency to conduct other business somewhere else. [16]

International Paper Company location. Chase Manhattan Bank in New York City loaned International Paper Company the money to build a new mill. The bank executives told IPC that they had loaned money to Kansas City Southern Railroad. Therefore, International Paper Company could build the paper mill anywhere they wanted to do so, but it had to be on a rail line that could use Kansas City Southern trains. William Buchanan had constructed the L & A Railroad from Stamps, Arkansas, through Springhill, Minden, and eventually into Vidalia, Louisiana. This satisfied the Chase Manhattan executives, so the mill was built along the L & A railroad tracks in Clifford, Louisiana (Clifford was later changed to Cullen). The Springhill location was chosen as the site for the new mill for several reasons, one of them being the location next to a railroad track that could use Kansas City Southern trains. Tons of raw material and manufactured paper were shipped in and out of the Springhill plant by Kansas City Southern trains.[17]

Pompey Flowers. The Springhill sawmill burned in 1912. William Buchanan decided to rebuild the mill bigger and better than the previous one. He brought into town from the Bodcaw mill at Stamps, Arkansas, an iron smith named Pompey Flowers. He and Dr. Rupert Butler played together when they were young boys. Pompey's mother had been a house slave from West Virginia who was brought to Bossier City by her master to a plantation called Palmetto (north of Bossier City). Pompey always wore a skull cap which he would tip when he met a "white brother." His blacksmith shop was located at the end of Giles Street across the railroad track from the sawmill. While he was working with iron to repair the mill, some of the sparks would fall into his boots. He would "dance a jig." The family did not know his age when he died, but Dr. Butler told them Pompey was 103 years old. His mind was still clear and his eyes had not dimmed.[18]

Rural food. Residents of Springhill had vegetable gardens to supplement their small wages at the sawmill. There were always plenty of vegetables, peanuts, popcorn, strawberries and ribbon cane syrup. Mayhaws along the creeks provided ingredients for jelly and marmalade. Hickory nuts, black walnuts, and pecans were abundant. Milk cows were owned by many families. The milk and churned butter would be lowered in a bucket into the cool well where it would be preserved during the summer heat. Families owned chickens, turkeys and geese. A chicken fried in hog lard served with biscuits and sawmill gravy was a favorite meal. Sawmill gravy was made by browning flour in fat after cooking chicken or steak. Water or milk was added to make the thickened gravy. Hogs were killed in the fall when it was cool. Smoked hams, salted bacon, and sausage were placed in the smokehouse for "curing." Red eye gravy was made by frying slices of ham and adding a small amount of water to the fat and natural juices of the ham.[19]

Rural living conditions. Food was cooked in cast iron skillets and dutch ovens on a wood stove. A good mattress could be made by filling the bed ticking with fresh straw. Feather mattresses were made from the feathers of ducks or geese. Empty feed sacks were sewn together to make sheets and dresses. Bed covers were quilts that had been made from scraps left over from sewing dresses. Soap was made from fat that was not used in cooking. Ashes from the fire were placed in a wooden barrel with a spigot in the bottom. Water was poured over the ashes which produced a liquid that drained out the bottom of the barrel. It was mixed with oil to make lye soap. After the mixture hardened, it was cut into bars.

Bathroom facilities consisted of a "John." Old Sears catalogues were placed in the John for paper. Baths were taken in a number three washtub or in the creek. Water was heated by placing a pan of water in the sunshine all day. In winter the water was warmed with a kettle of hot water that had been heated on the wood-burning stove.

Dirty clothes were placed in an iron wash pot with hot soapy water and boiled in the pot with firewood underneath it. Clothes were rubbed with a rub board, then rinsed twice in clear water. They were then hung on a clothesline or fence to dry.[20]

Wagons for school buses. Wagon drivers were added to the school system in Springhill in 1912. The horse-drawn wagons were the early "school buses" that carried the children to school.[21]

Lake Erling choir. Johnny Herrington and Johnnie Hill bought a small fishing camp on Lake Erling. A small group of men began to go to the camp to eat fish, have fellowship, and sing their favorite songs. The group grew so large that a room was added to the camp. When this location became too small, Ed Shultz invited them to his place. The group continued to grow, sing, and eat fish and steaks. They did not have an elected choirmaster, so they bought a jacket with "Ed Schultz, Choir Master" printed on it and designated that position to him.

Johnny Herrington, Johnnie Hill, Jack Byrd, Romeo Turgeon, and Bob Lavent; G. B. Pickett, A. B. "Shooter" Martin, Dwayne Wise, Bill Kottenbrook, and Marvin Tucker; Frank Milner, Syvell Burke, Joe Campbell, L. M. "Cotton" Cadenhead, and Jim Gayle; Billy Troquille, John Jackson, Avery Johnson and Ted Souter were participants.

One participant wrote, "There was a closeness among the men that made up the choir. Lifelong friendships developed. We were all very competitive men in the outside world, and that did happen at times, but for the most part, we left that behind when we were part of that special group called the Lake Erling Choir." [22]

Salesmen's wares displayed at O'Garte Hotel. An African-American man from the O'Garte Hotel located on Giles Street was called "Cotton Belt." He would meet the passenger trains traveling on the L & A Railroad, greet the salesmen, and carry their trunks of wares (clothing, cooking utensils, etc.) to the hotel for display. The salesmen would lay out their products for customers to view and buy. The hotel became the meeting place for commerce in the town.[23]

Ralph Ensey's bluejay contract. The blue jay is a beautiful bird, but it fights other birds. Blue jays were disturbing other birds in the lawn of Mr. Ralph Ensey. He contracted with the author's brother, Jerry Bonner, to kill blue jays for 25 cents per bird. It is not known how many blue jays were shot with a BB gun, but the contract was lucrative and the other birds in Mr. Ensey's lawn lived in peace.[24]

Tennyson's Drug Store move. Tennyson Drug store was located in Pine Woods Bank Building on the corner of Ensey and Giles Streets. They decided to move to the corner of Main and Church Streets, a location that became permanent for the drug store. When the time came to move, the employees pushed the merchandise in a wheelbarrow from the old location to the new location. It took several trips in the wheelbarrow to complete the move.[25]

Wheelbarrow ride down Main Street. When Shorty Fissel ran for Alderman on the City Council against Dr. Harris in 1951, Melvin Boucher made a bet with John Baker that Fissel would win. The payoff for the winner was a ride down Main Street in a wheelbarrow pushed by the loser of the bet. Fissel won. Baker lost the bet so he pushed Boucher from the City Park area to Branch Brothers Ford Agency in the wheelbarrow. Melvin Boucher was known as a "classy dresser," so he wore his suit and tie. Baker wore his khaki clothes. They had to stop and rest a few times before the journey was complete.[26]

Piccadilly Supper Club. A white frame building located on South Main Street at the present site of Crowe-Burlingame was the setting for the Piccadilly Supper Club. It was the "place to go" in Springhill from 1940–1955. Good steaks and dancing were the attraction for the evening.

It was first owned by June and Pearl Montgomery. In the late 1940s it was sold to George Mitchell who was a colorful entrepreneur. A patron said, "His idea of crowd control was the butt end of a pool cue." Fights were prevalent. The supper club was sold later to Mrs. Short.[27]

Princess Café hosts celebrities. The Princess Café was located near the corner of Main Street and Highway 157 East at the recent site of the R & R Vacuum Cleaner store. June and Pearl Montgomery were the owners. One story about a patron of the café could be truth or fiction, but it is told by older Springhill residents as fact. The infamous bank robber couple, Bonnie and Clyde, ate at the café, according to legend, immediately before they drove to Arcadia where they were killed by law enforcement officers.

Another story about a patron at the café is true. H. L. Hunt worked in the oil business in Cotton Valley before he became wealthy. On many occasions he ate meals at the café. In later years the name of café was changed to "The Ritz." [28]

Crow saves H. L. Hunt's life. Joe Crow and a team of rough necks were watching a "sick oil well" one night. A rumbling sound that shook the rig was heard from deep in the earth. The well began to "blow out." All of the men ran from the platform to a safe place. As Joe Crow passed a boiler on the rig, he saw a man asleep under it. He dragged the man off the platform to a safe place. This action saved his life because the rig was destroyed by the blowing oil. The man whose life was saved was H. L. Hunt.

Each Christmas Mr. Hunt, after he became wealthy from the oil business, presented a gift to Mr. Crow for saving his life. It was always an alligator billfold with $100 in it. Mr. Hunt also promoted Mr. Crow in his company through the years out of his appreciation for the man saving his life.[29]

Bootleg whiskey on the L & A Railroad. Two brothers from Springhill bought bootleg whiskey during prohibition days, rode the train from Springhill to Hope, and sold the whiskey to customers on the ride. They stayed overnight in the Barlow Hotel in Hope and returned to Springhill the next morning. All along the route they continued to sell their bootleg whiskey.

When the two brothers arrived in Springhill, they would buy more whiskey and make the same journey, selling along the way. There was a saying in Springhill, "You could go next door to get whiskey, but you had to go ten blocks to get a loaf of bread." [30]

Axe Café. Mr. and Mrs. Cordell Parker moved to the Arkansas state line in 1949. She built a restaurant next to their home in 1950. It was named the Axe Restaurant after the name of the Springhill High School newspaper. Later Liberty "Frenchie" Manuel built a new restaurant next to the Holiday Motel on South Arkansas Street. It became the new Axe Restaurant.[31]

Early Industrial Development funds. Henry Rhone was a successful Black business man in Cullen. He joined the Springhill/Cullen Chamber of Commerce in the 1960s. Long before the present Industrial District Park was planned and developed, Mr. Rhone presented the Chamber of Commerce a check for $1,500 for travel and incidental expenses incurred in the Chamber's effort to attract industry to Springhill. He was far-sighted about the need for industrial development in the city.[32]

First rodeo in Springhill. Robert McFadden produced the first rodeo in Springhill. I. B. "Boss" Slack provided the rodeo stock for the cowboys. There was no arena and no fenced area for the rodeo; the crowd formed a human circle to substitute for a fence. Its purpose was to keep the horses, calves, and bulls penned up for the rodeo. The site for the rodeo was near the present elementary school building next to Frank Anthony Park.[33]

J. F. Giles donates land for church buildings. In 1907 J. F. Giles, Manager of Pine Woods Lumber Company, donated one acre of land and all needed materials to the town to build a church building that would be used by all denominations. Since Baptists and Methodists were the two major denominations at that time, they worshipped on alternate Sundays. The two groups met in the union building until 1920.

In 1918 the Methodists began construction on a sanctuary. It was located on land donated by Mr. Giles through Pine Woods Lumber Company. It was constructed on the site of the present sanctuary. This happened after the Baptists bought the union church building from the Methodists for $500.

In 1927 the New Bethel African American Methodist Episcopal Church building was destroyed by a storm. It was rebuilt by the generous gifts of three ladies and the gift of lumber by Mr. Giles.

Three churches were constructed through J. F. Giles' generous gifts of lumber and land.[34]

Dr. Butler's black bag. Dr. Rupert Butler was a beloved physician who practiced medicine in Springhill in the early days of the city's history. He would not refuse anyone a house call day or night. His medicines were carried in a black satchel as he made his rounds on horseback, in his automobile or on foot. One distinct image of Dr. Butler is from a photograph of him walking to his office in the snow wearing an overcoat, his black hat, and carrying his black bag. His granddaughter Flo married Dr. Wayne Sessions, a respected surgeon in Springhill, and they now possess this valuable treasure of early medicine.[35]

W. W. Monroe's painting of Uncle George's Café hangs in City Hall.

John Baker and Melvin Boucher—
the famous wheelbarrow ride down Main.

Picadilly Supper Club on east side of south Main, about 1944.

Chapter 19

Biographies of Citizens

Springhill is a small town in North Louisiana that has been blessed with citizens who have demonstrated outstanding leadership in institutions. Others have gained national prominence through their talents and personal achievements. These two factors—leadership or national prominence—have been the benchmarks for choosing the following biographical vignettes. Many citizens who have made contributions to Springhill through faithful service have not been included because of space limitations, although they are worthy of recognition.

It is remarkable that so many people from a small town in a small state have achieved national fame. A list of those who have achieved national acclaim are: William "Billy" Baucum, coach of the Louisiana State champions in football and basketball; Jesse Boucher, National Realtor of the Year; John David Crow, Heisman Trophy winner; Eludia Flanakin, President of the American Legion Women's Auxiliary; Linda Baucum Goldsberry, Miss Louisiana and participant in Miss America contest; Michael Haynes, television star and advertising model for the Winchester Man; Sherry Boucher Lytle, Sugar Bowl Queen; Dr. Donald Mack, outstanding pediatrician and nationally recognized supporter of St. Jude's Children's Hospital; Joe Stampley, country music "rock and roll" singing star; Robert Charles Smith, National Commander of the American Legion; Dr. Samuel Williamson, President of the University of the South at Sewanee and history professor at West Point; D. C. Wimberly, National Commander of Ex-Prisoners of War; and John Stephens, National Football League's Rookie of the Year in 1988.

These biographical vignettes are presented to focus on the historical significance of individuals to Springhill. Hopefully, opening a small window to their lives will inspire readers to volunteer for service to the community and to seek excellence in their own lives.

Biographies

Clary Anthony. Clary Anthony is a third generation sawmill chief executive officer. His grandfather, John Anthony, was a postmaster and sawmill entrepreneur. He began the sawmill business for the Anthony family in Southwest Arkansas.

Clary Anthony was reared in Hopeville, Arkansas, on a farm. There were six brothers and two sisters in the family. After he completed high school in Hopeville, he entered Texas Wesleyan College. During this time World War II began, so he enlisted in the United States Navy to participate in flight training. After he completed his obligation in the military, he and his brother Melvin came in 1946 to Springhill to purchase Frost Lumber Industries sawmill and timberland.

In 1952 he assumed management responsibilities for the mill operation. The mill was owned by Anthony Forest Products, Inc., a partnership composed of Beryl, Bruce, Clary, Frank, and Garland Anthony. He served as chief executive officer of the mill until it was closed in 1972. During this time he was chief executive officer of Anthony Forest Products, Inc.

Clary Anthony, three brothers, and Bill Bryan purchased O'Bier Insurance Company in Springhill. At a later time, he bought the insurance company from the brothers and operated it for many years as Anthony-Bryan Insurance Company. He sold it to Mac Pace.

Mr. Anthony entered the banking business in Citizens Bank and Trust Company as a stockholder. In 1982 he became Chairman of the Board. Since that time he has led the bank to expand its services to Plain Dealing and Minden. He is responsible for the attractive mural on the East side of the bank building.

He served on the city council 1958–1962 during the administration of Jesse Boucher. It was during this time that many improvements were made in the city.

He continues to serve on the board of directors of Anthony Forest Products, Inc., and Chairman of the Board of Citizens Bank and Trust Company. He is especially proud of the new Frank Anthony Memorial Park named after his father, Frank Anthony.

He described Springhill as a fine place to live and raise a family. He said, "I have been here since 1946 and I don't plan to leave willingly."

Esther "Pat" Bailey. "This community has been wonderful to me and I have tried to give back to it as my way of service," said Pat Bailey. This native of North Carolina has demonstrated this philosophy numerous times. Before her business was expanded into other cities, she attended every funeral and visited with every family personally prior to the funeral.

Pat was reared in Princeton, North Carolina, as one of twelve children to tobacco farming parents. Chores were assigned to all, which meant rising as early as 4:00 A.M. to accomplish the assigned tasks before attending school.

After she graduated from high school, she left her Quaker home and earned a degree in interior design from Gifford College in North Carolina. She met her husband, the late R. A. Bailey, on a bridge in the moonlight while he was on leave from military service. They married and had three daughters.

The couple moved to Springhill in 1957. They purchased the funeral homes owned by George T. Norton Corporation. These included homes in Springhill, Plain Dealing, and Cotton Valley. In later years they purchased homes in El Dorado, Arkansas; Junction City, Arkansas; and Magnolia, Arkansas. Today Mrs. Bailey owns seven funeral homes, three in Arkansas and four in Louisiana. She has twenty-four employees.

Pat Bailey is a member of the Rotary Club and Senior Friends. She has been involved in community projects and organizations, enjoys water color and china painting, and participates in one of her favorite activities—bridge.

In her business life and in her private life she practices a principle taught by her parents: treat everyone the way you like to be treated. She said, "My profession requires you to be on call twenty-four hours a day, seven days a week. It's a very demanding profession. You've got to be there for others."

Ed Bankhead. This Springhill City Councilman has served seven terms since 1974. He was born January 26, 1942, during the war years. These early years were spent in Plain Dealing, Louisiana, where he attended Bossier Parish High School. In 1960 Mr. Bankhead became a citizen of Springhill. During his first two years in the town he worked for Anthony Forest Products, Inc. in the sawmill.

He became an entrepreneur in 1962 when he developed his own construction company. Later he added a lawn service to his business enterprise. Much of his time today is spent supervising the Camelot housing district in Springhill. He is a member of the Harrison Chapel church where he serves as a deacon. He has been a Christian since 1951.

Mr. Bankhead reflected on his City Council work and his life in Springhill when he was asked, "What do you think of our town?" He responded, "I love Springhill. It is in my heart. Springhill holds its own.

Population has decreased since 1979, but jobs are available. If we had affordable housing, people would live within the city rather than working here and living out of town. We need to increase the population with more housing. I'm proud of Springhill. It is in pretty good shape. The Mayor is doing a good job."

William "Billy" Baucum. Coach Billy Baucum was the son of Oscar Fulton Baucum and Delia Lizabeth Slack Baucum of Webster Parish. His great-grandfather, Shelby Baucum, was from North Carolina and migrated to Webster Parish in the mid-1800s. After his mother died he was raised by his aunt and uncle, Grace and Roy Tapscott.

Baucum attended the public schools in Springhill. In 1932 he graduated from Springhill High School. During his student days he played quarterback for the Lumberjacks football team and forward for the basketball team. He received an athletic scholarship to Normal State College (Northwestern State University) where he excelled in academics, student activities, football, basketball and track.

In extra-curricula activities he was captain of the football team 1935–1936; president of the "N" Club 1934–1936; president of the senior class 1935–1936; member of the school yearbook "Potpourri" and student council 1934–1936. He was drafted into the U. S. Army where he attained the rank of captain. He was wounded in the invasion of Normandy.

After graduation he began his coaching career at Shongaloo High School. Later he was backfield coach at Springhill High School. When he was promoted to head football coach at Springhill, he also coached basketball and track. He was successful in all three sports. He is the only coach in Louisiana to have teams win the state championship in both football and basketball. His 1952 Lumberjacks were state champions and his 1953 basketball team won the state championship. One of his outstanding athletes, Jack Montgomery, said Coach Baucum was responsible for the winning attitude at Springhill High School. He gave a speech to the student body at a time when the spirit of the school was low. He used illustrations from his war years to inspire the students to develop a team spirit, a winning attitude, and a discipline for the task.

His coaching skills were recognized by the other coaches of the state. He was elected to coach the North Louisiana All Stars and he was elected president of the Louisiana High School Coaches Association. Many of his high school athletes played on Southwest and Southeastern Conference teams. He coached John David Crow, a future Heisman Trophy winner.

After Coach Baucum retired, he was principal of Browning Elementary School in Springhill. Then he became a supervisor for the Webster Parish Schools. He said, "An honest man is the noblest work of God."

Tylon Blanton. As Tylon handed the author some materials I had requested about his life, he walked from my front lawn toward his automobile, turned around and said with a big smile, "I love Webster Parish." That love has been demonstrated by a life committed to various and numerous causes that have helped the people of the Parish.

He is a lifetime resident of Springhill. During his high school days he was a superb athlete—an award winning halfback in football and a track star. He won the Ed Shultz Award in 1952 as the outstanding back on the team. After he fulfilled his obligation in the military, he went into the automobile business in 1955 and later into the chemical distribution business. In 1968 he ran for Police Jury which launched him on a career in Webster Parish government. He has been elected for seven terms of four years each. He was President of the Police Jury for eight years 1977–1984. He is presently serving another term.

During his tenure on the Jury Mr. Blanton has been a member of the Federal Land Use Board with Russell Long to review the wet lands laws. He has been elected to the Region 4 Police Jury Association that represents ten parishes. This group presents resolutions to the state legislature to equalize funds for North Louisiana, avoid doubling the excise tax that penalizes our area, and lobbies for funds for the criminal justice system, police equipment, and roads in the area. Their interest is helping North Louisiana. He has been the driving force behind building a new Butler Health Unit on Church Street.

Blanton was instrumental in creating North Webster Parish Industrial District and receiving grants for the Industrial Park which seeks new industry for the area. He is Vice-Chairman of the Bi-State Corridor Commission for Highway 371, a member of the Board of Directors of Northwest Louisiana Coordinating and Development Council, and member of the Advisory Board for the City of Springhill.

He is a charter member of I-69 Transcontinental Highway Coalition. As a member of the Coalition he was instrumental in having the corridor near Springhill designated as the route. He is working now to get I–69 as close to Springhill as possible. In his campaign literature he says, "I will continue to work for a better future for our area, always with the best interest of our citizens at heart." His record of getting the job done substantiates that promise.

Jesse Boucher. They call him "Springhill's Mr. Business Man," "A mover and shaker of the city's progressive spirit," "Mr. Northwestern State Athlete," and "Mr. Real Estate Developer." All of these names are accurate.

Jesse Boucher is a native son of Springhill, graduating from Springhill High School in 1931. He was an outstanding end for the Lumberjacks which earned him an athletic scholarship to Northwestern State University. While he attended Northwestern he was a member of the football and track teams for four years. He ran on the Conference winning mile relay team and received honorable mention in football.

In addition to his athletic abilities, Mr. Boucher was highly regarded by his classmates for his leadership qualities. He was Senior Class President and a member of the Student Government Council. He also excelled in academics.

Following his graduation from Northwestern, he taught and coached at Doyline High School for three years. He joined the Naval Air Force in 1941. He resigned from the Naval Air Force and returned to Springhill to organize an insurance agency with Wilburn Slack his long-time friend. He was called back into the Air Force during World War II.

When he returned to Springhill following his discharge from the Air Force, he continued in the insurance business. He also opened a construction supply business in 1946 known as B & S Supply. This launched his career in housing developments throughout Arkansas and Louisiana. In Springhill he built the sub-divisions of North Acres, Morelane, Whiteway, Marie Street, Meadow Creek, Frost sub-division, and B & S Park. He built sub-divisions in other cities including Bossier City, Baton Rouge, and Natchitoches, Louisiana, and in Little Rock and Pine Bluff, Arkansas.

He has held the following leadership positions and received the following honors: Outstanding Young Man of the Year, Who's Who in the National Register, Builder of the Year in 1949, Chairman of the American Legion Post, President of Springhill Lion's Club, P. T. A. and Springhill High School Alumni Association. He has been Mayor of Springhill. He is a charter member of the SHS Alumni Association and an active member of the Bossier City and Shreveport Chambers of Commerce.

Jesse Boucher said of his early ambitions, "It seemed I was always struggling for something I didn't have. It's never really been about the money. It is about making people happy."

Annie Lee Branch. Annie Lee Branch celebrated her 100th birthday on March 20, 2001. She was born in Timpson, Texas, the daughter of Bill and Lacy Bateman. She was married to the late J. B. Branch, Sr., known in Springhill as "Little Branch."

She and her husband joined his brother, J. Alford Branch, known in Springhill as "Big Branch," to operate the Ford dealership. At that time it was owned by J. F. Giles. In 1934 she and her husband purchased the Ford dealership with his brother J. A. Branch as a partner. They renamed the agency Branch Brothers Ford Company.

Her family and her church have always been the important priorities in her life. She also enjoys quilting, sewing and playing skip-bo. She continues to teach others to quilt and crochet. In times past Mrs. Branch said she enjoyed lawn work and cooking. She has been active in the Springhill United Methodist Church for 75 years. She also enjoys activities with the Sew and Sews group and Senior Friends.

Mrs. Branch described her life by saying, "As my children grew up and had children of their own, my life changed. I had time to enjoy sewing, quilting, and playing skip-bo. I love my children and my many friends. I adore the fact that I have been lucky enough to watch my great-great grandchildren grow up. I have seen many changes in my 100 years, but nothing compares with the joy that I have felt watching my family grow."

She reflected on her life in Springhill and said, "I am happy that Blance and I chose Springhill as our home. I couldn't imagine spending my life anywhere else without all my wonderful friends."

John Marvin Browning. John Browning, the youngest of his family, was born in 1888 in Haynesville. After his first wife died, he married Lillian Oden. The couple lived in Springhill many years.

He was a charter member of Murray Memorial Church, now known as Central Baptist Church. He was a banker and a merchant in Springhill. John Browning served as Chairman of the Board of Springhill Bank and Trust Company. His merchantile store was located across the street from Tennyson Drug Store.

One citizen remembers him as a very religious man and an influential man in Springhill.

Merit Taylor "Mitt" Browning. Mitt Browning was born on a farm near Haynesville in Claiborne Parish in 1875. He married Pearl Nations. The couple settled near Springhill.

In 1914 Dr. Melvin Browning, Dr. Joe Browning, and Mr. W. R. Oakley founded a general merchandise store on "The Hill," the area on Butler Street and Highway 157 near the present-day Brookshire store. At that time Springhill business establishments were on "The Hill." The shopping section was a small cluster of buildings on that site. In a few years Mr. Oakley sold his interest in the store to Mitt Browning. In 1923 W. H. Oden and John Browning acquired the business. Mitt Browning established his own merchantile store near the present-day Butler property. In 1924 he moved his business to Main Street where it remained until his retirement.

His community activities were many, including organizing the American Legion baseball team for high school boys, membership in the Masonic Lodge, and membership in the Welcome Church. He made the first deposit in Citizens Bank and Trust Company by bringing his money to the bank in $20 bills in a gallon bucket.

Mitt Browning served on the Webster Parish School Board from 1916 to 1950. He was elected President of the Board. When he died in 1957 the school board designated the name of the new elementary school in Springhill as "M. T. Browning Elementary School."

William Buchanan. He was a Tennessee boy from Decherd who migrated to Arkansas in 1869 to earn his fortune in sawmills. When he arrived in northern Arkansas, he had with him a portable sawmill. In 1873 he moved to a small village called Texarkana where he met Joseph Ferguson, a wealthy landowner in the village. His friendship with Ferguson was probably the deciding factor in stopping his southern and western migration. It was here that he went into the sawmill business three miles from the Sulphur River.

After experiencing setbacks in the sawmill business, Buchanan traveled to Leadville, Colorado, to seek his fortune in silver mining. This business venture did not work out, so he returned to Texarkana in 1880. It was here at this time that his business activities became successful. In 1899 he gained control of Bodcaw Lumber Company. Then he began to expand his sawmill business into Louisiana. Eventually he owned seven sawmills and the L & A Railroad.

In 1896 Buchanan built a Bodcaw Lumber Company in a small village on the Arkansas state line called Barefoot, Louisiana. This village became Springhill, Louisiana. The sawmill was in the center of the business life and activities of the town.

He owned the sawmill, the houses, and all the buildings in town although he never lived in Springhill.

In addition to building the sawmill in Springhill, he also built the L & A Railroad from Stamps, Arkansas, to Vidalia, Louisiana. This railroad line, which ran through Springhill, transported his manufactured lumber to markets throughout the United States. As he expanded his railroad, he either built or influenced greatly seven towns along the route. Springhill was one of them.

Springhill cannot call him a local citizen, but the town was a direct result of William Buchanan's sawmill and railroad between 1896 and 1912.

Rupert Butler, M.D. The doctor with the "little black bag" is a description of Rupert Butler's medical practice. He carried his medicines with him in the black bag, made house calls, walked or rode his car to patients' homes at all hours of the day and night, and sometimes spent the entire night by the bedside of a critically ill patient. He was born October 25, 1879 in Mars Hill, Arkansas and died in 1948.

Rupert Butler attended Memphis Medical School for twelve years to train for the medical profession. When he married Katherine Goodwin in Red Land in 1907, the couple made their home across the road from the Salem Baptist Church. Pine Woods Lumber Company asked him to be the mill doctor in 1919. He moved his family to Springhill where he and Kate bought a lot on "The Hill" for their new home. The street in front of the house would later be named Butler Street in their honor.

His daughter said Dr. Butler's primary purpose in his life was to practice medicine and to take care of his family. He was on call every day of the year. Those who knew him said he would never refuse to go on a house call day or night when he was called. Because of his devotion to helping others through his medical practice, he was loved, respected, and admired. Local citizens paid him a tribute by naming the Webster Parish Health Center in his honor—"Butler Memorial Health Center."

Careece Perritt, a public health nurse who worked for him, said in 1946, "He had compassion, energy, devotion, and concern for others." The plaque on the Butler Memorial Health Center reads, "In memory of Dr. Rupert Butler who unselfishly served the community for 31 years."

Malcolm C. Colvin. Malcolm Colvin began a 62 year career in the automotive industry in 1937 in Troup, Texas. In 1945 he moved to Springhill and became the General Motors dealer for 54 years. He was a successful and loyal General Motors dealer through hard work and commitment to customer service.

Mr. Colvin attended Haynesville High School and Tyler Business College. He served on the Board of Directors for Springhill Bank and Trust Company and was Chairman of the Board for several years.

He was a member of the Lions Club in Springhill and an active member of Central Baptist Church. Mr. Colvin and his wife of 62 years, Floy DeLoach Colvin, were active in the programs and development of North Webster Parish through their participation in community and civic activities.

John David Crow. Second place would never satisfy John David. He had great desire to be a winner. After a difficult childhood, he put on a football uniform in the seventh grade. It was the beginning of countless honors, sweet victories, and the Heisman Trophy.

He was born in a farmhouse north of Marion in Union Parish just off the Huttig highway. He went to the Springhill elementary and high schools. In 1952 he led the Lumberjacks to win the state championship. He was equally good as a basketball player, leading the Lumberjack basketball team to win the state championship in 1953. In one of the greatest football games in Louisiana history he led the Lumberjacks to tie Byrd High School 20–20 in 1953.

After graduating from Springhill High School, he played football for Bear Bryant at Texas A & M where in 1957 he won the Heisman Trophy. In the same year he led the Aggies to a national championship. After he graduated with a degree in Business Administration he served as Assistant Athletic Director at Texas A & M where he is credited with moving A & M into the forefront of collegiate athletics, especially in the area of gender equity.

Crow played 11 years in the National Football League for the Chicago Cardinals, St. Louis Cardinals, and San Francisco Forty-Niners. He served as Assistant Coach under Bear Bryant at Alabama and head coach at Northeast State University in Louisiana. He is now a business man in Texas.

Richard J. Cullen. Richard Cullen was not a citizen of Springhill, but he had tremendous influence on the destiny of the town. He was employed by Riordin Paper Mills, Limited, when he was a young mechanical engineer. At a later date he was assigned to construct paper mills for International Paper Company at Merritton, Ontario; Bastrop, Louisiana; Camden, Arkansas; Mobile, Alabama; Panama City, Florida; Georgetown, South Carolina; and Springhill, Louisiana. His excellent work created a good reputation in the company. He was elected a Director of International Paper Company in 1936, President a few weeks later, and Chairman of the Corporation in 1943.

It is said of Cullen that he has possibly done more than any other man to further the development of paper mills in the South. He was the first man to open up the vast resources of southern woodlands for the raw material to make kraft paper. Later in his career he pushed back the technological frontier in the South by recognizing the possibilities of mass producing kraft container board on the Fourdrinier machines. For over thirty years he was one of the major figures responsible for the development of an industry which has been of incalculable value to the social and economic well-being of the South.

Cullen was President of International Paper Company when the Springhill paper mill was conceived and constructed. He was Chairman of the Board in 1943 when the Springhill mill expanded production at a rapid pace.

The town of Clifford located across the state highway from the Springhill mill changed its name to Cullen in honor of Richard J. Cullen. Although he was not a citizen of Springhill, he had great influence on the growth of the town through his leadership as President and as Chairman of the Board of International Paper Company.

Travis Farrar. Coach Farrar was a native of Magnolia, Arkansas. He was an outstanding athlete in football and baseball at Magnolia High School and Southern Arkansas University. When he pitched for the American Legion baseball team in Magnolia and for Southern Arkansas University, he led his team to win the AIC championship in 1954 and 1956. He was known for his blazing fast ball. When he played quarterback for Southern Arkansas University, the team won the AIC championship in 1953.

He coached at Springhill High School for 38 years. Thirty of those years he was head coach of the Lumberjacks. His coaching record was 254 wins, 89 losses, and 13 ties. His teams had 18 playoff appearances. They won 12 district championships and one state championship. He was inducted into the Louisiana High School Hall of Fame and he was selected as the Louisiana High School Coach of the Year in 1960.

Coach Farrar died at half time during a Lumberjack football game. The Webster Parish School Board named the football stadium after him and Billy Baucum. It is known today as the Baucum-Farrar Lumberjack Stadium.

Eluida Flanakin. "Mrs. Flan," as she is affectionately called by many of her students, is a native of Pittsburg, Mississippi. Her grandfather was Ellis Hoffpauier from Germany. She lived with her grandparents until she was thirteen years of age.

In 1928 she entered Bellhaven College in Jackson, Mississippi, with anticipation of becoming a missionary. Here she studied the Bible, church history, and world missions which expanded her view of the world. After she attended Bellhaven for a brief time, Mrs. Flan visited Houma, Louisiana, where she taught French and the first grade, and did volunteer work at the Clanton Memorial Methodist Church. For a brief time she taught in the Shongaloo school system.

In 1940 she graduated from Northwestern State University. Following graduation she began her teaching career in Springhill where she taught the third grade and freshman English. Her teaching career was interrupted briefly in 1942 when she married Bernice Flanakin and moved to Mineral Wells, Texas, while he served in the military. In 1945 she began teaching once again in Springhill. The subjects she taught were Spanish, junior literature, and senior literature. She said her philosophy of teaching was, "Never pretend you know something when you don't. Look it up with the students and learn together." It was during this time that she began a twenty-five year sponsorship of the Student Council.

In 1972 Mrs. Flanakin began working with the American Legion Auxiliary which opened doors of opportunity for her as alternate delegate to the national convention, chaplain, historian, and President of the Fourth District. In 1979—1980 she was elected as President of the Louisiana American Legion Auxiliary. In 1980 she served on the National Executive Committee.

Mrs. Flanakin is a member of the Springhill United Methodist Church where she has taught Sunday School for fifty years. She has one son, four granddaughters, and two great grandchildren.

John D. Herrington. John Herrington has been the Mayor of Springhill 1978–1986 and 1994 to the present. Many accomplishments have occurred during his tenure. These include, but are not limited to, building the new airport, Frank Anthony Memorial Park, developing the Main Street program, encouraging the work of the Beautification Committee, and initiating the Centennial Celebration.

He moved with his family from Bastrop, Louisiana, in 1937. In Springhill he developed a food service business in candy and tobacco vending machines. He lived also in Branson, Missouri, where he was in the entertainment, hotel, and land development business.

Mayor Herrington has been an active citizen in various organizations. He was president of the Springhill Bank and Trust Company, an alderman on the City Council for two terms, a member of the Springhill Lions Club, the Masonic Lodge, the Shrine

Club, and a member of the Co-ordination and Development Corporation. He was a leading proponent of the North Webster Parish Industrial District Board that encourages new industries to locate in Springhill. One citizen said of the Mayor, "Johnny keeps on keeping on."

Dr. A. C. Higginbotham. The years 1943 and 1944 were life-changing years for A. C. Higginbotham. This farm boy from near Bienville, Louisiana, finished high school, drove a taxi for M. C. Burnham, worked in the Burnham drug store, and married his wife, Maxine.

After high school he worked for the L & A Railroad one year. Mr. Burnham from Cullen offered him two things: a job and a loan to go to pharmacy school. He and Maxine moved to New Orleans in 1944 to enter Loyola University School of Pharmacy. His academic career was interrupted by military service in Japan. After he was discharged in 1947, he returned to Loyola to complete his pharmacy degree. While he was working in Henderson, Texas, as a pharmacist, he received an invitation from Jesse Boucher to return to Springhill and operate a drug store. In 1950 he purchased one-half interest in the drug store which became known as H & B Drug. In 1962 he purchased the other one-half interest in the drug store. He had the pharmacy, a gift shop, and a soda fountain until he discontinued the soda fountain in 1969. Mr. Higginbotham remarked that he had the first "wall to wall carpeted store in Springhill." He returned to Loyola University in 1982 to earn his Doctor of Pharmacy degree.

He has been active in religious, civic, business, professional, and political life of Springhill. He is a deacon and active member of Central Baptist Church, a member of the Lions Club where he served as President, a charter member of the local Chamber of Commerce, a member of the Masonic Lodge and the American Legion. He was the organizer of Medisave Pharmacy in Springhill. In 2000 A. C. Higginbotham was asked to serve as Chairman of the Centennial Celebration Committee to plan the Centennial Celebration in February 2002.

Sherry Boucher Lytle. Sherry, the daughter of Jesse and Eloise Boucher, graduated from Springhill High School in1963. She was actively involved in numerous school organizations and events: Student Council, cheerleader, Math Club, and the girls basketball team. She was high school beauty four years.

Sherry was Sugar Bowl Queen in 1965 and Holiday In Dixie Queen in 1966 while she was a student at Northwestern State University. She went under contract with Universal Studios in Los Angeles, entered the University of Southern California where she earned a Bachelor of Arts degree in speech and drama and became an actress.

Her career included leading roles in the "Ozzie and Harriet" show, *Couple of White Chicks Sitting Around Talking, Shepherd of the Hills* with Richard Arlen, *White Lightning* with Burt Reynolds, *Five Days From Home* with George Peppard, and *Eating With Francis Bergen and Kathleen Crosby.*

In recent years she has been a real estate broker in Bossier City, Louisiana, President of the Women's Council of Realtors, a member of the Louisiana Realty Political Action Group, a member of the Screen Actors Guild, and the American Federation of Television and Radio Artists. She is an organizer of Mothers Against Guns in the Classroom and is in demand as a motivational speaker.

R. O. Machen, Sr. When he became Superintendent of Webster Parish Schools, R. O. Machen challenged the school district to restore the schools to a "level of recognized quality." An evaluation team from George Peabody College for Teachers in Nashville, Tennessee, researched, evaluated, and made recommendations regarding every area of the Webster Parish School District. His administration and others throughout the years have used these guidelines to improve education in the school district and to "restore the schools to a level of recognized quality."

His life experiences prepared him for this leadership role. He was born in Winnfield, Louisiana. He moved to Springhill when he was 10 years old. After he graduated from Springhill High School, he earned his B.A. and M.A. degrees from Louisiana State University.

His 43-year career in education began in Webster Parish in 1926 at Evergreen High School. After serving as principal of Weston High School near Jonesboro (1929–1932), he returned to Webster Parish where he served as principal of Doyline High School, Cotton Valley High School, and Springhill High School.

In 1952 Mr. Machen was apponted Supervisor of Instruction where he served until 1961. He was elected Superintendent of Webster Parish Schools in 1961 and held that position until his death in 1969.

He was named Educator of the Year in 1964 by the Minden Chamber of Commerce. Machen was a member of Phi Delta Kappa, an honorary organization for men in the field of education, a member of the Downtown Lions Club and the First United Methodist Church in Minden.

Dr. Donald G. Mack, Sr. Donald Mack, M.D., is a 1947 graduate of Springhill High School. He is the son of Mr. and Mrs. Willie Mack, long-time business people in Springhill. Dr. Mack earned his Bachelor of Science degree at Centenary College in 1951 and his medical degree from Louisiana State University in 1956 where he specialized in Pediatrics. After he served in the military in the medical corps, he began practicing medicine in Shreveport in 1961.

He is a member of eight medical organizations: Shreveport Medical Society, Louisiana State Medical Society, Louisiana State Pediatrics Society, American Academy of Pediatrics, American Academy of Allergy, American Medical Association, American Board of Pediatrics, and Northwest Louisiana Pediatric Society. Currently he serves on several hospital staffs in Shreveport and Bossier City: Christus Schumpert Medical Center, LSU Medical Center, Willis-Knighton Medical Center, and Christus Schumpert-Bossier Medical Center. He is Associate Clinical Professor of Pediatrics at LSU School of Medicine in Shreveport.

Dr. Mack was instrumental in sending the first patient from out-of-state to St. Jude Medical Research Center in Memphis. He continues to support the Research Center by promoting an annual Dream House give away in the community. Proceeds from the ticket sales go to St. Jude Medical Research Center.

He received the prestigious Jefferson Award for his medical work with children and his service in the city of Shreveport.

Willie Mack. During World War I Lebanon was under Turkish rule and under an embargo by the French military. Life was very hard. Real famine was present in the country. When the war was over, Willie Mack migrated to the United States at the age of 21 to the city of Marshall, Texas. He and his uncle went into business together in

Shreveport, Louisiana. This Lebanese from Birbara began to carve out a career in the dry goods department store business. He moved to Springhill from Longview, Texas, in 1940 because of the influence of his best friend, Mr. Griffin. In Springhill he established a department store which was located in three places at three different times on Main Street. He also participated in the oil business in the area. He was a very successful merchant until he retired in 1968.

During his career in Springhill he was active in the Chamber of Commerce, the Lions Club, the Country Club, Club 40 Business Men's Fellowship, and the Springhill Presbyterian Church where he served as an Elder. In 1973 he moved to Shreveport where he and his wife Mary lived until his death in 1981. They were the parents of five successful children. His daughter Dorothy said, "He really loved Springhill. He hated to move away to Shreveport."

Charles McConnell. Few men in Springhill have earned the title, "Mr. Citizen." Charles McConnell is one of them. His formal education was obtained at Rayville High School, Northeast Louisiana State University in Monroe, and Louisiana State University Law School. After he graduated from law school in 1950, he began practicing law with his brother in Springhill.

He is a veteran of World War II. For twenty-four months he was in the U. S. Coast Guard amphibious forces which participated in the campaigns in North Africa, Sicily, Italy, and Normandy. He was active in Springhill organizations which included the Springhill Community Council, American Legion, Lions Club, Quarterback Club, Citizens Council, Masonic Lodge and Chamber of Commerce. He served as president of several organizations.

Charles McConnell was elected Mayor in 1954. During his term as Mayor numerous contributions were made to the city. These included mail service to every home, renaming of all streets, daily bus service from Shreveport, street construction, radio equipment for police cars, extension of water and sewer lines, and the development of municipal planning program. He was elected to the Webster Parish School Board in 1961. Later Charles McConnell served as President of the Chamber of Commerce in Springhill.

One of McConnell's most significant accomplishments was the formation of Citizens Bank and Trust Company. He and Waymon Oden organized the bank in 1955. For many years he served as chairman of the board.

He was a member of the Springhill Methodist Church where he served on the official board and taught a Sunday School class. Later in his life he was a member of Central Baptist Church where his reputation grew as a knowledgeable Bible student and teacher of a men's Sunday School class.

Andrew Jackson McDonald, M.D. Dr. McDonald was born near Emerson, Arkansas, on December 20, 1870. His father, John Cox McDonald, and three uncles migrated from North Carolina.

Dr. McDonald was the first licensed graduate doctor to serve in Springhill. In 1900 he graduated from the University of Arkansas Medical School. However, he had practiced several years in Shongaloo with Dr. Purnel Burns and in Minden before he graduated.

While practicing medicine with Dr. Bond in Shongaloo in 1898, Mr. Harris from Pine Woods Lumber Company invited Dr. McDonald to come to Springhill to provide medical care for the employees. He agreed to come twice a week. He rode his horse from Shongaloo to Springhill on those two days and gave medical attention to the Pine Woods Lumber Company employees and families. He continued this routine until he moved to Springhill in 1900. At that time he opened an office for medical practice and a merchandise store for business purposes.

On two occasions he took a leave of absence to study surgery. In 1904 he attended Washington School of Medicine in St. Louis. In 1914 he attended New York City Polyclinic for post-graduate study in surgery.

In 1915 Louisiana passed a law that would not allow doctors to practice in the state if they received their medical training outside the state. However, they were allowed to live in adjacent states and practice medicine in Louisiana. His lawyer advised him to move one mile north of Springhill into Arkansas where he had several hundred acres of land. He made this move and continued to serve as a physician in both Louisiana and Arkansas. He died in 1961 after practicing medicine for 50 years.

John M. "Jack" Montgomery. Law was his destiny, but before law, Jack Montgomery was an outstanding athlete. Beginning in the third grade he did not miss a football game at home or away. He was a mascot cheerleader in the fourth and fifth grades and manager of the football team in the sixth and seventh grades.

In high school he was involved in numerous extra-curricular activities: Student Council, Pelican Boys State, S-Club which he organized, and class president. He played tight end on the 1952 championship football team and was all-district when he played on the 1953 championship basketball team.

Following graduation Jack attended Tulane University where he lettered in football. After he was graduated he attended Louisiana State University Law School where he received his Juris Doctorate degree in 1962. After he served in the Judge Advocate Corps in the Air Force, he returned to Springhill to practice law.

He was President of the Lions Club, Chamber of Commerce, and Little Boys Baseball League. He was selected Outstanding Young Man in Bossier Parish in 1969. In 1968 Mr. Montgomery was elected to the Louisiana State Senate. He also served one year as Judge of the Minden City/Ward One Court. He resides in Minden where he practices law.

Henry Rhone. Henry Rhone is an example of the fulfillment of the American dream. He journeyed from plowing fields on a farm to wealth in grocery and furniture sales, real estate and cattle.

He was the youngest Rhone among 13 brothers and sisters in a family from Natchez, Mississippi. Mr. Rhone was raised on a farm near Jonesville, Louisiana by his grandmother who could neither read nor write. He learned to plow when he was eight years old. His early school was limited to three months per year in an African-American school that met in a church building. During the other months he worked on the farm.

The work ethic was strong in his life. Also, he saved part of his small income in these early years. He sent back home twenty-five dollars out of the thirty dollars he made per month when he worked for the Civilian Conservation Corps during the Depression years. On rainy days when the survey crew could not work, he cut hair in the CCC camp. By the time he resigned from the CCC he had saved $500, a large sum of money during the Depression days. Mr. Rhone said, "When I went to town, I saw a man wearing black pants, a white shirt and shined shoes. He did not farm. I decided I wanted to do something like that."

He opened a store in a ten by twenty-foot space in Cullen, Louisiana, where he sold sugar cane, peanuts, milk, and soda pop. He raised turnip greens and sweet potatoes that he sold in the store. After a few years in the grocery business, he added furniture and hardware to his inventory. He would trade groceries for gasoline stamps, food stamps, and sugar stamps during World War II. As his business grew, he invested in real estate and cattle.

His life went from a farm man, to a grocery man, to a construction supply man, to a real estate man, to a cattleman. He lived conservatively, saved his money, and invested wisely. He said he met his wife when he offered her a slice of watermelon during one of her visits to her brother's house in Cullen. After they married, she taught him high school correspondence courses for 12 years until he completed his high school work.

Henry Rhone is an example of an African-American man who lived in the segregated South in the early part of the 20th century, survived the Depression, and became wealthy by conservative living and regularly saving and investing his money. He has lived the American dream.

Erling Riis. Mr. Riis was born in Christiania, Norway, where he received his education in Norwegian schools. He graduated in 1909 as a mechanical engineer from Kristiania Technical School. After he graduated he worked for various paper mills in Norway. In 1911 he came to the United States. His first job was that of a shepherd in Montana. He later worked in paper mills in Wisconsin where he attained the position of pulpwood foreman.

In 1915 Mr. Riis came to Bogalusa, Louisiana, to work in the Fibre Board Company as their chief engineer. On September 1, 1920, he moved to Bastrop, Louisiana, where he was in charge of construction for the Louisiana mill. He had other successful jobs as chief engineer in Camden, Arkansas; Mobile, Alabama; Panama City, Florida; Georgetown, South Carolina; and Springhill, Louisiana mills.

In May 1942 he was elected Vice President of International Paper Company where he had full responsibility for all production, research and development, and construction. Mr. Riis was considered one of the foremost engineers in the paper industry. He developed many labor-saving devices, processes, and products, and he was considered an authority in efficient construction and operation of paper mills. He was chief engineer of construction at the Springhill mill in 1937. In 1956 the company named Lake Erling after him.

Ed Shultz. A contemporary leader called him "Mr. Springhill." This active citizen was a successful businessman, a mayor for twelve years, an alderman for twelve years, and president of Springhill Bank and Trust Company. He was born on July 11, 1904, in Oklahoma Territory. He attended school at Prague Oklahoma High School. His first job was with Champion Oil. During this time he was informed by his sister that International Paper Company planned to build a paper mill in Springhill, Louisiana. He wanted to arrive before the masses of people came to town, so he bought the telephone exchange in Springhill to make his living. He owned the telephone exchange from 1937 to 1952 when he began working for the Springhill Bank and Trust

Company. In 1958 he became president of the bank, a position he held until his death on March 23, 1970.

Ed Shultz was mayor for three terms 1940–1952. When he did not run for a fourth term due to illness, he got out of the political arena four years. He was elected alderman for three terms 1956–1968. Springhill was in its greatest period of growth during the time he served as mayor and as alderman. Accomplishments were massive. Because he had a talent for organizational skills, he began many new projects for the city. He organized the Shrine Club and the Methodist Men's Sunday School class. He worked as mayor and alderman to build the new Civic Center, pave Main Street, install lights on the streets, revise the city's bookkeeping system, shut down open saloons, expand the fire department, get firearms off the streets and into the hands of policemen only, and revised the bond program which saved thousands of dollars for the city. He and other business leaders organized the Chamber of Commerce. The old city hall, the swimming pool, and the fire station were built during his tenure as Mayor. He and the other aldermen initiated and encouraged the police jury to build Butler Memorial Health Unit.

Mr. Shultz was interested in the welfare of the youth of the city. His extreme pride in the Lumberjack football team was demonstrated when he had small cards printed which read, "If you play from your heart, you can win." He handed them to the team as they ran onto the field of play for the state championship.

His philosophy of life was centered around his Christian faith and two principles by which he lived: "Never argue politics and religion," and "There is always bad in the best people and there is good in everybody."

Jerry Wayne Sessions, M.D. Dr. Sessions is a respected surgeon in Springhill. He is qualified to practice in any hospital in the United States, but he prefers to use his enormous skills to heal the sick in his hometown.

He was an excellent athlete in high school. He excelled in football as an all-state quarterback, in basketball as an all-district player, in baseball as an all-district player, and as tennis and track star. He won a football scholarship to Louisiana State University. As an LSU Tiger he played in the Cotton Bowl against Arkansas in 1966.

His medical training began when he was awarded the LSU Alumni Scholarship to medical school. At Louisiana State University Medical School he earned his M.D. degree which was followed by a residency in general surgery at LSU-Shreveport. He won the Paul Abramson Award for the outstanding third year resident in 1975. He was the chief surgery resident in 1976, the assistant clinical instructor in 1976, and the clinical instructor in surgery at LSU-Shreveport in 1976–1980.

Dr. Sessions is a member of the American Board of Surgery, the American Medical Association, the Louisiana State Medical Society, the LSU-Shreveport Surgical Society, the Society for Laparoendoscopic Surgeons, and the Southern Medical Society.

In Springhill he is the Lumberjack team physician, a member of the Lions Club, Quarterback Club, and Webster Parish Medical Society. He has served as chief of staff at the Springhill Medical Center. Since 1976 his professional practice has been at Doctor's Clinic in Springhill. He is a member of Central Baptist Church in Springhill.

Ira Benton "I.B." Slack. He was born in 1890. His great grandfather, James Slack, came from Kentucky to fight in the War of 1812 in New Orleans. After the war James Slack met Hugh Coyle from Ireland and married his daughter. The family moved to

Mississippi, then to Louisiana near Cotton Valley and Shongaloo. The purchased land in the area south and east of the present-day International Paper Company.

I.B. Slack helped develop Springhill Bank and Trust Company. He was on the board of directors, serving as chairman of the board several years.

O.M. "Boss" Slack. O. M. Slack was the first real estate developer in Springhill. He was born in 1882 in Webster Parish and died in Springhill in 1968. Mr. Slack developed sub-divisions on his land in the early days of Springhill. Housing was unavailable when the paper mill was constructed in 1937 which caused families to live in tents in the city park. At this time he built houses in Eastside, Whiteway, and near Howell Elementary School. He gave the land to Webster Parish School District to build Howell Elementary School. He also built the Spring Theatre, Springhill Bank and Trust Company, and part of the city block where Hull Furniture is located.

Mr. Slack helped other Springhill citizens begin their business careers in the city by providing jobs and training in all phases of his construction business.

R. A. "Buck" Smith. An outstanding citizen said about R. A. Smith, "He was one of the finest men I have ever met. When he told you something, you could count on it. He was a man true to his word." People called him "Mr. Springhill."

Robert Alonzo Smith was born in Walton County, Georgia, on April 18, 1879, one of 14 children. In December 1880 the Smith family moved to Warren, Arkansas, where Mr. Smith lived until he was a young man. He attended Poughkeepsie Business School in Poughkeepsie, New York, after he completed high school.

In 1909 R.A. Smith came to Springhill as a timekeeper for Pine Woods Lumber Company. In 1918 he was made secretary-treasurer of the company. At the death of J. F. Giles he became vice-president and manager of Pine Woods Lumber Company. He worked in this position until 1936 when Frost Lumber Company bought the mill. Frost Lumber Company asked him to be the superintendent of Frost Industries where he served until 1946.

He guided the growth and development of Springhill for many years as he served on the city council, built an ice plant, began the water department, and organized the electric and gas company. He also had an insurance company during this time.

When he observed virgin timber being cut out of this country, he realized Springhill would need new industry to survive. He had the vision and faith to work with Harvey Couch of the L & A Railroad, E. A. Frost of Frost Lumber Industries, and United Gas Company to seek new industry. They developed the Webster Development Corporation for this purpose. This organization bought land and sought International Paper Company to locate a plant in Springhill. It was primarily through the efforts of Mr. Smith and the Webster Development Corporation that International Paper Company located the kraft paper mill in the town. Much of the prosperity and well being of this city can be attributed directly to his leadership.

He was a steward of the Springhill United Methodist Church and taught a *Bible* class for forty years. He served as a trustee and chairman of the board of trustees for the church. The beautiful sanctuary and educational building were constructed under his leadership.

R. A. Smith was described as a friendly man with no idle words and no pretense. He talked straight to the point, was a man of his word, and possessed great faith in God. In his lifetime in the town he was truly "Mr. Springhill."

Robert Charles Smith. Robert Charles Smith is the son of Mr. and Mrs. R.A. Smith. His father managed Frost Lumber Industries in Springhill and assisted International Paper Company in buying options for the land used for building the paper mill. His family contributed much to the schools, the city, and the Springhill First United Methodist Church.

He is a 1935 graduate of Springhill High School and a graduate of Louisiana Tech with a degree in commerce and accounting. Louisiana Tech once selected him as Alumnus of the Year.

Mr. Smith served in the Air Force where he attained the rank of Chief Warrant Officer. After his discharge from the Air Force, he returned to work at International Paper Company. He was director of financial administration and chief financial officer.

His civic responsibilities included President of the Lions Club, the Springhill Chamber of Commerce, and the Louisiana Tech Alumni Association. He was a member of the Louisiana Welfare Board for twenty-three years and was chairman of the board for eight years.

In 1948 he was selected by the Springhill Jaycees as the "Young Man of the Year." He was involved in Boy Scout work for which he received the Silver Beaver award in 1973.

His most notable work was with the American Legion. In 1977 the American Legion organization elected him national commander. In this office he represented the Legion in all fifty states, Korea, Taiwan, Central America, France, Germany, and Flanders Field. Often he met with the President of the United States and spoke on national television in behalf of the veterans, widows, and orphans. Commander Smith would introduce himself to Legion groups all over the world by saying, "I'm from Louisiana and I live in Springhill."

William Byrd Smith. The Springhill school became a state approved high school under the direction of W.B. Smith in 1909. He served as principal of the school 1909–1916. He was principal of the Cotton Valley school 1916–1921 and principal of the Sarepta school 1921–1926. Due to his wife's illness, he resigned as principal in 1926 and returned to the classroom in Springhill. He retired from teaching in 1937.

Mr. Smith was the first degreed teacher in Webster Parish. He won a scholarship to Peabody College in Nashville, Tennessee. In 1903 he graduated from the college and began his teaching career. He conducted the first graduating class in 1910-1911 with two graduates—his daughter Bertha Smith Reynolds and W. A. Miller.

When he retired from teaching after 28 years, the Webster Parish School Board sent him a letter congratulating him on his years of service by stating, "We appreciate your long term of service; it has been characterized by such splendid traits as an excellent attitude, fine cooperation, and superior work. Of what more could anyone boast? Your untiring service will long be remembered as a faithful servant of the public."

Joe Stampley. Joe Stampley graduated from Springhill High School in 1961. He became a national celebrity by bridging the gap between rock "n" roll, rhythm and blues, and country music. He pioneered a style known as "new country" a decade before this marketing niche had been given a name.

His climb to fame has paralleled his climb in the record charts. In the 1960s Joe was the lead singer for a pop-rock group, The Uniques. While he played for the group, they had two hit songs, "Not Too Long Ago" and "All These Things."

In 1971 Joe signed with ABC-Dot for which he recorded seven albums. He did a remake of "All These Things" which skyrocketed to the number one record on the charts. In 1975 he signed a contract with Epic Records for whom he recorded 13 albums. He also had a string of hits with his label mate Moe Bandy. They recorded a take-off on Boy George, "Where's The Dress," which won the American Video Association's award for Video of the Year in 1984. In 1980 Joe and Moe were recognized as the Country Music Association's Vocal Duo of the Year.

Joe Stampley has over sixty charted records to his credit which ranks him 30th in Radio and Records Twenty Years of Excellence magazine. Joel Whitburn's Billboard Top Country Singles ranks him 52nd among all country artists from 1944–1993 for charted singles.

He has a high energy style that often involves the audience in sing-alongs, hand clapping, and dancing in the aisles. He is currently on tour playing concert halls, casinos, clubs, and festivals. From humble beginnings this country rock star has carried the name "Springhill" to national prominence and recognition.

Jimmy Thomas. Jimmy Thomas is serving his third term on the Webster Parish Police Jury. He was elected in 1992. This native of Shongaloo was born March 10, 1939. He moved to Springhill in 1956 after being raised in the neighboring town.

Through the years he has worked at various stores and industries including Polk Furniture, Nations Brothers Packing Company, the United States Government Ammunition Plant, Anthony Forest Sawmill, and International Paper Container Division. At the present time he is owner of J. & M. Construction Company.

He is a member of Ephraim Masonic Lodge #224 and Harrison Chapel Baptist Church where he serves as a deacon. When he was asked what he thought of Springhill, he replied, "It is a nice place to live. However, the town does need more industry for jobs. My wife, Melba, and I have raised our children here. I like the environment." He said of his police jury work, "There are good days and bad days, but things usually work out for good. The office gives me an opportunity to help people. I enjoy that."

Dr. Samuel Ruthven Williamson. Sam Williamson was a 1954 honor graduate of Springhill High School. He was valedictorian of his graduating class, a participant in extra-curricula activities, and an academic scholar.

Mr. Williamson earned his B.A. degree from Tulane University in 1958, an A.M. degree in history from Harvard, and a Ph.D. degree from Harvard University in 1966. Sam was a Fulbright Scholar at the University of Edinburgh 1968-69. He received an honorary degree from Virginia Theological Seminary. He studied also at Harvard Graduate School of Business Administration in its Advanced Management Program.

Dr. Williamson taught history at the United States Military Academy, West Point, and at Harvard University, University of North Carolina at Chapel Hill, and the University of the South in Sewanee, Tennessee. He has served in numerous administrative capacities including Assistant Dean of Harvard College, Director of Curriculum in Peace, War, and Defense, Dean of the College of Arts and Sciences in the General College at Harvard, and Provost and Chief Academic Officer at the University of North Carolina at Chapel Hill. He has served on seventeen consortiums in the American Political Science Association and as a director on boards of many organizations and institutions. Dr. Williamson is a prolific writer of book reviews and journal articles and of books related to defense strategy and history.

He served as Vice Chancelor of the University of the South in Sewanee, Tennessee 1988–2000. He continues writing and teaching in retirement.

Jonathan Washington. He was born in 1932 in the Plainfield Community near Emerson, Arkansas. During his early years he attended Doss Town Elementary School, a Methodist Minister's school. Religion has always been a part of his life. He was raised by his grandmother when his mother died in 1942. During these years he attended the Washington Church of God in Christ. His favorite memory is sitting around the dinner table in the evening quoting Bible verses with the family. Each child was required to quote a verse before the meal was served.

The Rev. Washington was involved with Webster Parish public schools for more than thirty-four years—Shongaloo, Brown High School, Brown Middle School, and Brown Junior High School. He taught civics and business administration most of those years. Also, he was principal of Brown Middle School for twenty-one years. During this tenure he organized an intramural program for all the children and led the school to upgrade test scores to the national level.

He has pastored churches for thirty-two years in the African Methodist Episcopal denomination. Currently, he is the presiding elder of the churches. He supervises fourteen of them in the A.M.E. denomination. One of his proudest accomplishments is the establishment of the Springhill Community Center owned and operated by the Jonathan Washington Evangelistic Association. He serves as the current President of the NAACP.

Eugene Waters. He loved antique cars but his life was vitally contemporary as a community leader in Springhill. Early childhood and youth were spent in Pittsburg, Texas.

Mr. Waters and his wife, Virginia, made their journey to Springhill in 1941 where he became manager of the State Theatre. After he served his country in the Pacific Theatre in World War II, he returned to Springhill to continue his job as manager of the theatre. He said, "Business was good. There was nothing else to do in Springhill so people would fill the theatre when it opened."

A Director of the Springhill Bank and Trust Company Board asked him to work for the bank. The year was 1947. He worked there for over forty years until his retirement in 1988. He served as President of the bank for fifteen years. He retired in 1988 to pursue his personal interest in automobiles. During his career he was a member of the Louisiana Bankers Association and traveled extensively in the United States and abroad during this time.

Mr. Waters was an active leader in the North Webster Industrial District. He served on the original committee that formed the District. He was a coordinator in the Salvation Army, an officer in the Springhill Lions Club, and a faithful member of Central Baptist Church.

His hobby was collecting and restoring old automobiles, driving them to conventions, and driving them in parades. He was a member of the Antique Car Clubs of America and the Classic Car Clubs of America. He died September 22, 2000.

Janis Stroud Willis. She is called one of Springhill's most dynamic promoters. At the present time she manages the Main Street Program which renovates the storefronts of businesses with grants from the state.

Jan is a native of Springhill who graduated from Springhill High School in 1955. Through continuing education courses she has prepared for her career by studying accounting, insurance, notary public laws, computers, real estate, and motivational sales. Her career includes work with Anthony-Bryan Insurance and O'Bier Insurance as bookkeeper, co-ordinator, and insurance sales; Kenyan Enterprises as Director of Accounting Operations and Secretary-Treasurer of four corporations; Sikes Construction Company as bookkeeper; and Main Street Program Manager for Springhill. Her civic work has been diverse and effective: Director on the Chamber of Commerce Board, fundraiser for Frank Anthony Park, Director of the Lumberjack Festival for two years, and a member of the Beautification Committee in Springhill. She said, "Doing volunteer work for the city of Springhill has been the most satisfying to me."

Her accomplishments are outstanding: President and director of the Chamber of Commerce, organizer of the Visitor Information Center, and chairman of the Lumberjack Festival. She has led businesses to beautify store fronts, organized the Mardi Gras parade, organized the Main Street Classic Car Show, and promoted the Main Street from Minden to Springhill program.

Jan is optimistic about the future of Springhill. She works diligently to secure grants for the Main Street Program and other events which draw others to the city. She said of her job as Main Street manager, "As an open-ended position this is one of the most rewarding challenges I've ever had."

Wilbur "Will" Wilson. "It speaks of America," he says, "that the farm boy from Louisiana eventually served ten years as the highest administrative officer in the world's largest service organization."

It all began when Will was born near the Shiloh community west of Springhill. He was raised on a forty-acre farm six miles west of Plain Dealing. After he finished Plain Dealing High School, the sixteen-year-old youth set out for Chillicothe, Missouri, to enroll in business college. He waited tables and performed other chores to pay for tuition, room and board at the college.

Upon graduation Mr. Wilson set out for Chicago to find a job. It was the Depression years, so jobs were difficult to find. After searching for thirty-seven days as one of thousands looking for a job, he found one in the young International Association of Lions Clubs. He was hired to process records at the Lions International headquarters. During the next forty six and one half years he rose in the Lions organization from records clerk to executive administrator over 1.2 million members and 32,000 clubs. During his administration he worked hard to expand the Lions eye care and blindness prevention programs and in the formation of Lions clubs for women.

He has received more than twenty International presidential awards and he has received the Ambassador of Good Will award, the highest award a Lion can receive. He retired in 1978 after ten years as executive administrator and more than forty-six years as a dedicated Lion worker.

Mr. Wilson retired in Springhill with his wife Jeanne and daughter Susan. "Seems to me, and I know it was that way when I left here, in small town areas people have more concern for each other," said Will. The Lions Club of Springhill is honored to have him as a member.

D. C. Wimberly. D. C. Wimberly is an American hero. He is a small-town boy who became National Commander of Ex-Prisoners of War. His boyhood was spent in Ringgold, Louisiana. When he graduated from high school he earned his bachelor's degree from Northwestern State University and his master's degree in education from Louisiana State University.

He entered the teaching profession at Bienville, Louisiana. Later he entered the administrative division of education by serving as principal of Shongaloo High School, Springhill Junior High School, and Browning Elementary School in Springhill.

During World War II he was drafted into the 101st Infantry of the 3rd Army. His military training was completed at Fort Hood, Texas. When he was in Europe, his platoon searched out German units that were assigned to advance movements of the panzer divisions. During this time he was captured by German soldiers while he was on patrol (See Chapter 15 for the complete story of his capture). He was a prisoner of war in Germany from November 25, 1944 to May 15, 1945. He was liberated by the Russians on April 22, 1945, and returned home to Louisiana on June 29, 1945. He received an honorable discharge December 12, 1945.

After he returned home, he became involved in a young organization called "Ex-Prisoners of War." In 1974–1975 he was elected National Commander of the Ex-Prisoners of War. He continued his work as teacher and principal in the public schools of Webster Parish, specifically in Springhill. During his retirement he continues to be involved in Ex-Prisoner of War events. He and his wife live in Springhill.

PART SEVEN

Reflecting on the Forest— Past and Future

Tommy W. Duke's young pine forest on Hwy. 157 East.

Pine trees, the renewable resource, being grown by Sim McDonald on Hwy. 157 East.

Chapter 20

Historical Conclusions

When one approaches the end of a book, conclusions become inevitable. The desire to draw generalizations from the specific events and individual personalities tempts the writer to summarize his specific research into over-arching themes. But Will Durant, who wrote ten monumental volumes of *The Story of Civilization*, warns that, "The historian always over-simplifies, and hastily selects a manageable minority of facts and faces out of a crowd of souls and events whose multitudinous complexity he can never quite embrace or comprehend." [1] Yet we attempt conclusions.

The following list of "historical conclusions" has been compiled by the author from opinions expressed by citizens during the writing of this book. They do not have primary or secondary references. They are totally the thoughts of the author gleaned from conversations with many people. Thus, the responsibility for the list is the author's alone.

There are ten conclusions related to the city of Springhill. The *Historical Conclusions* are summaries of events that have affected the history of the city. Some are obvious while others are not so easy to observe and generalize.

First, Springhill has been built on the economy of the pine tree and the oil well. Since William Buchanan constructed Bodcaw Lumber Company in the village of Barefoot in 1896, the forest has been at the heart of growth. International Paper Company was constructed in 1937 which propelled the town into its period of greatest growth.

Second, since 1979 when International Paper Company closed the production section of the plant, the leadership of Springhill has attempted to diversify the business activity. Small businesses have moved into the North Webster Parish Industrial District buildings. The leadership of the city and of the Industrial District continues to work toward bringing new industry into the area. A major difficulty with this goal is the lack of an interstate highway or a railroad line running through the city.

Third, Springhill is thought of as an industrial town. The sawmill and the paper mill created this image. Today, business and industry have expanded to include oil and gas activity, and the manufacture of plywood, home grills, corrugated fiber boxes, and mobile offices.

Fourth, the history of Springhill has been characterized by feelings of hope and despair. Global events such as World War II and the Great Depression and local events such as the closing of the sawmill and paper mill have created despair. Yet people express great hope for better days in the midst of economic downturns. A spirit of optimism is present among the citizens.

Fifth, each time that adversity comes to the town, people of Springhill have demonstrated that they have courage, vision, and tough-mindedness. These characteristics have been a source of strength for growth in our competitive society. The town has always bounced back after adversity.

Sixth, there is a spirit of renewal in the people as they prepare to celebrate one-hundred years of history since the town was incorporated. Hope has replaced despair, rebuilding has replaced deterioration, and progress has replaced inertia. The Main Street program which was begun in 1997 has been a driving force for restoration and revitalization.

Seventh, the city is becoming a retirement center. Persons who were reared here are returning when they retire. Those who have remained in Springhill through the years to work are choosing to keep their permanent residence in the city. People are attracted to the beauty, recreation facilities, medical services, and small-town atmosphere of the area.

Eighth, the quality of life in this small industrial town has been enriched by an excellent library, art league, civic clubs, welfare services, youth activities, recreational opportunities, schools, and churches of various denominations. These organizations and activities have inspired a talented group of volunteers to give their time and financial resources to sustain this quality of life.

Ninth, ethnic groups are learning to live in harmony and goodwill with each other. Both African-Americans and Anglos participate in school programs, police jury service, city council, festivals, and recreation. Both groups are working on problems of integration and adequate housing. Racial harmony has been the result of these efforts.

Tenth, after difficult years following the closing of the production division of International Paper Company, the city is regaining pride in its existence. There is a renewed interest in the beautification and history of Springhill.

Summary

These ten historical conclusions are not inclusive ideas of all those expressed by citizens. However, they provide a basic framework to consider the unfinished business that, if prioritized and acted upon, will thrust the city into a new era of growth and quality of life development. They are presented as items to encourage creative thinking about the future of Springhill.

Chapter 21

Unfinished Business

The Challenge

Growing a city is a dynamic, not a static, enterprise. We will pause to reflect on our rich heritage, but there remains much to be done. Again, Will Durant gives us insight into an unfinished agenda. He wrote, "If we put the problem further back, and ask what determines whether a challenge will or will not be met, the answer is that this depends upon the presence or absence of initiative and creative individuals with clarity of mind and energy of will (which is almost a definition of genius), capable of effective responses to new situations (which is almost a definition of intelligence)."[1]

First, it is imperative to seek diversified small businesses rather than depend upon one industry for economic stability. Many individuals need to work together aggressively to reach this common goal. Large industry is welcome, but diversity in business is the objective.

Second, an effort should be made to tap into the talent of senior adults and retirees for volunteer work in various organizations of the city. Likewise, an emphasis should be given to meeting needs of this group in housing, recreation, and meaningful activities. For example, building a retirement center needs to be a priority.

Third, although many persons are involved in activities and organizations of Springhill, there needs to be an organized effort to broaden the base of volunteers. Organizations will be stronger and the culture of the city will be enriched.

Fourth, quality education and adequate facilities in our schools should be a common goal for all citizens, but especially for parents of school-age children. Likewise, school administrators and teachers should continue their efforts to provide the best possible learning experiences for each child. It is a task that is never completed.

Fifth, progress in race relations should be a priority in attitude and action in the city if harmony is to continue among different ethnic groups. A pro-active stance rather than a re-active stance builds good relationships. The continuation of racial harmony is a vital need for community stability.

Sixth, encourage youth to assume leadership roles and to have happy life experiences during their growing years. It is a responsibility and a privilege to provide good role models for our young people in the areas of government, career, life-style and religious faith. Every citizen can assume the role of encourager.

Seventh, cultural activities and institutions can improve the quality of life among Springhill citizens through drama, art, sports, recreation, and intellectual challenges. A drama theatre and a permanent Museum of History are two worthy goals for the city. The Historical Commission is very capable of operating the museum and keeping the history of the city before the people.

Eighth, it is important for citizens to promote and practice good communication among each other. The use of new technology, newspapers, public forums, television, and the internet are available for this cause. Mrs. S. R. Williamson, Sr., a long-time resident of Springhill, explained how important communication was in the early days of growth and development. She wrote,

> Today we hear a great deal about the importance of communication. With all the media available now we should stop to realize how important communication was in 1938 and the ensuing years. Springhill was fortunate from the start when Floyd Barnes was editor and publisher of the *Springhill Press and News Journal*. Springhill was doubly fortunate when the Garrison family (Doyle and Avis) moved from Virginia and became editors and publishers of the *Springhill Press and News Journal*. And now in 1987 the tradition is continued through Danny and Ann Scott and Danette Scott. Another essential means of communication was the telephone service of Ed Shultz and his family who moved from Oklahoma.[2]

Ninth, teach the history of Springhill to citizens through periodic emphases. Knowledge of events and personalities of the past will inspire and enrich the lives of all. It will especially lead the youth to appreciate their heritage and community.

Tenth, encourage citizens to be faithful in church attendance and recognize God as the source of all good things in life. Springhill has strong churches of various denominations that provide a foundation for moral and spiritual values. This must continue in the midst of our secular society.

These items of Unfinished Business ideas can be translated into priorities, goals, and programs for the city as its history unfolds during the next one-hundred years.

Conclusion

The shining "city on the hill" built on the economy of the pine tree can justify a celebration of one-hundred years of history. However, the journey has just begun! After we have paused to participate in the Centennial Celebration, we must look forward toward the road ahead. Robert Frost captured the gravity of the journey when he wrote:

> The woods are dark and deep
> And I have miles to go before I sleep.
> And miles to go before I sleep.[3]

The author has no doubt, whatsoever, that one hundred years from now another writer will pen a history of the bicentennial of Springhill. Perhaps he will conclude his book, as this author does, by writing, "Well done ordinary citizens who have accomplished extra-ordinary deeds. You are a courageous and visionary people."

Appendices

Appendix A
People of Northwest Louisiana

PRE-HISTORY (10,500 B.C.–800 B.C.)
- Paleo Indian . . . . 10,500 B.C.
- Meso Indian . . . . 6,000 B.C.
- Neo Indian . . . . 2,000 B.C.
- Pre-Caddo. . . . . 8,000 B.C.

CADDO INDIANS (800 B.C.–1543 B.C.)

EUROPEANS (1543 A.D.–1803 A.D.)
- DeSoto . . . . 1543 A.D.
- LaSalle . . . . 1682 A.D.
- Slaves . . . . 1717 A.D.

AMERICANS (1803 A.D.–2002 A.D.)
- Louisiana Purchase . . . . 1803 A.D.
- Statehood . . . . 1812 A.D.
- Buchanan. . . . . 1896 A.D.
- Centennial. . . . . 2002 A.D.

Appendix B
Major Eras of Springhill History

PRE-HISTORY (10,500 B.C.–800 B.C.)
- Natural Resources
- Native Americans . . . . 10,500 B.C.
- Pre-Caddo. . . . . 8,000 B.C.

CADDO INDIANS (800 A.D.–1543 A.D.)

EUROPEANS (1543 A.D.–1803 A.D.)
- Pre-Caddo. . . . . 8,000 B.C.
- Caddo . . . . 800 A.D.

EUROPEAN EXPLORERS (1543 A.D.–1803 A.D.)
- DeSoto . . . . 1543 A.D.
- LaSalle . . . . 1682 A.D.
- SLAVERY . . . . 1717 A.D.

AMERICAN INFLUENCE (1803 A.D.–1896 A.D.)
- Louisiana Purchase . . . . 1803 A.D.
- Statehood . . . . 1812 A.D.
- Buchanan. . . . . 1896 A.D.

CITY OF SPRINGHILL (1896 A.D.–2002 A.D.)
- Early Settlers . . . . 1810 A.D.
- Bodcaw Lumber Co. . . . . 1896 A.D.
- Barefoot To Springhill. . . . . . 1896 A.D.
- Pine Woods Lumber Co.. . . . 1897 A.D.
- Incorporation . . . . 1902 A.D.
- Early Growth . . . . 1896 A.D.–1937 A.D.
- Rapid Development 1937 A.D.–1979 A.D.
- Economic Loss. . . 1979 A.D.–1997 A.D.
- Rebuilding City . . 1997 A.D.–2002 A.D.

Appendix C
Buchanan's Empire

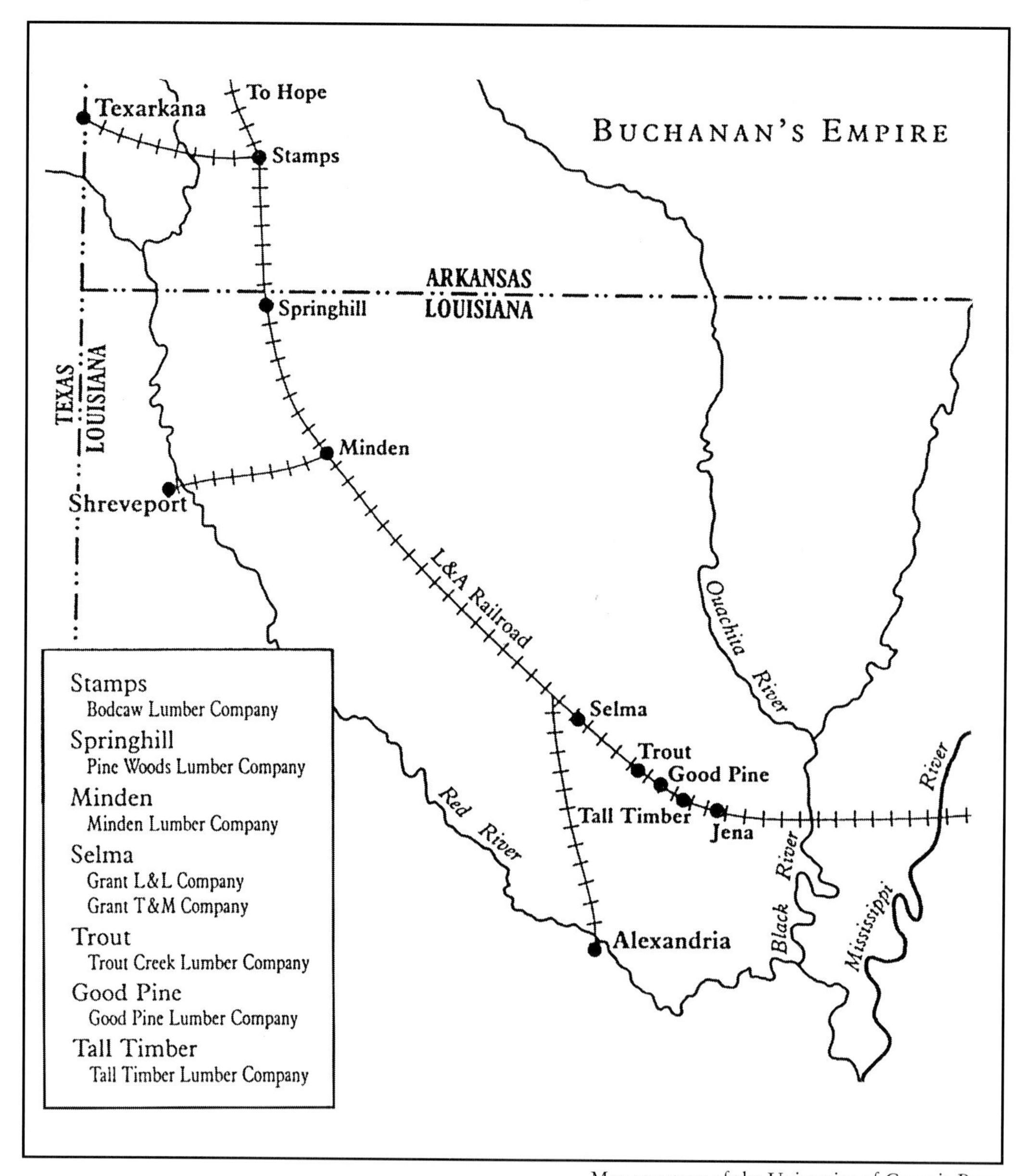

—Map courtesy of the University of Georgia Press

Appendix D
Historical Events in Springhill

10,500 B.C.	Native Americans in area
800	Caddo Indians in area
1803	Louisiana Purchase—Louisiana became part of United States of America
1811	Isaac Alden first English-speaking settler in North Louisiana
1817	Colonel William Clark set off wave of migration from Middle Tennessee to Northwestern Louisiana
1824	Fires clear trackless wilderness in North Louisiana
1871	Webster Parish organized out of Claiborne Parish
1896	William Buchanan built Bodcaw Lumber Co. sawmill in Barefoot, Louisiana
1896	Barefoot Station name changed to Springhill by J.F. Giles
1897	Bodcaw Lumber Co. changed name to Pine Woods Lumber Company
1897	L & A Railroad came to Springhill
1897	First School in Springhill
1901	L & A Railroad incorporated
1902	Springhill incorporated as city
1912	Pine Woods Lumber Co. burned and rebuilt
1916	Bank of Springhill organized
1921	Oil discovered in Haynesville & Shongaloo
1933	Pine Woods Lumber Company closed
1936	Frost Lumber Company bought Pine Woods Lumber Company
1937	International Paper Company began operations
1938	Dr. Garnea builds first hospital— Swan Clinic
1943	Springhill Bank & Trust Company founded
1946	Anthony Forest Products, Inc. bought Frost Lumber Industries
1950	Youth Recreation Center built
1957	Robert Charles Smith elected National Commander of American Legion
1957	John Crow wins Heisman Trophy— best football player in nation
1959	Springhill Medical Center constructed; Linda Baucum wins Miss Louisiana contest
1969	Civic Center and City Hall complex constructed
1970	Integration of public schools
1971	Webster Parish Centennial
1972	Anthony Forest Products sawmill closed
1974	Anthony sawmill burned
1974	North Webster Industrial Board organized
1979	International Paper Company closed
1983	Lumberjack Festival begins
1997	Main Street Program begins
2002	Springhill Centennial Celebration of City's Incorporation

Appendix E
Mayors of Springhill

Will Morris 1902–1914
Walter Modisette 1914–1925
Tom Haynes 1925–1929
Walter Mallock 1929–1933
Marvin Dillon, M.D. 1933–1937
W. E. Coyle 1937–1938
C. J. Provost 1938–1940
Merrell Cox 1940–1942
Ed Shultz 1942–1954
Charles McConnell 1954–1958
Jesse Boucher 1959–1962
James Allen 1962–1974
M. A. Gleason 1974–1978
Johnny Herrington 1978–1986
Curtis Smith 1986–1994
Johnny Herrington 1994–

Appendix F
Principals of Springhill High School

W. A. Miller 1895–1901
Arthur Pope 1901–1905
W. F. Mothershed 1905–1909
William Byrd Smith 1909–1915
Rev. Lee 1915–1917
Captain Smith 1917–1921
J. L. Liggen 1921–1927
S. R. Emmons 1927–1934
J. L. Cathcart 1934–1940
R. O. Machen 1940–1952
E. O. Cooper 1952–1968
Ed Olive 1968–1969
Harold Bartlett 1969–1971
Ray Burnham 1971–1986
Wayne King 1986–1999
Nathan Gills 1999–2001
Mary H. King 2001–

Appendix G
Founding of Springhill Churches

Central Baptist 1922
Church of Christ 1941
Dorcheat Acres Baptist Church 1967
Eastside Baptist Church 1953
First Assembly of God 1945
First Baptist Church 1902
Harrison Chapel 1905
New Bethel AME 1921
North Arkansas Street
  Church of Christ 1972
Sacred Heart Catholic Church 1949
Southern Methodist Church 1968
Springhill Christian Church 1973
Springhill Missionary
  Baptist Church 1952
Springhill Presbyterian Church 1939
Springhill United Methodist
  Church 1895
Temple Baptist Church 1960
Trinity Worship Center 1984
Walnut Road Baptist Church 1956
Washington Church of God
  In Christ 1947

Appendix H
Organizations

Ark-La Partners in Genealogy
Art League
Church Organizations
Civic Club
Habitat for Humanity
Lions Club
Lumberjack Festival Association
Masonic Lodge
Quarterback Club
Riding Club
Rodeo Association
Rotary Club
Senior Friends
Springhill Country Club
Webster Parish Council on Aging

Appendix I
Cultural and Recreational Events

Miss Lumberjack Contest	February
Mardi Gras Parade	February
Tribute to America	July
PRCA Rodeo and Parade	August
Lumberjack Festival and Parade	October
Art League Arts and Crafts Show	November
Christmas Parade	December
Churches' Christmas Programs	December
Sportsman's Paradise	Daily
Organized Sports Youth and Adults	Weekly
Recreation Center	June–August
Drama Theatre	Pending

Appendix J
Springhill Veterans of World War II[1]

John Wayne Adams
James Allen
Howard Alexander
Alton L. Andrews
Lonnie Andrews
Tommie C. Arnold
Cecil Bail
Sullivan William Baker
William Baker
Robert L. Barnes
Sam Baxter
Billy Baucum
Scedell Bearden
Marvin Beavers
William T. Beavers
L. E. Berry
Harry Blanchard
Jesse Boucher
Melvin Boucher
James Branch
Edward Branson
William T. Bowen
Burton B. Boyette
Thomas R. Brady
John Robert Braley
Wilton Braley
Web P. Braley
Edward Branson
Zebedee Branton
Frank Brinkman
John M. Browning
James Burns
Pleasant N. Burns
Susie Lynn Burns
Willis Burns
Charles Burnham
Drayton C. Burrell
Ed Burrell
John D. Burrell
Joe Campbell
James Canterbury
James Carmack
Willie Carroll
John T. Carter
Adair Cason
Charles Cason
Johnnie Cason
Percy Charles
Waldo Chastant
Willie Cheatham
Elbert Cochran
Maxwell Cochran
Melbourne Cochran
Harold Coleman
Roger Coleman
William Ross Cooper
Bob L. Coyle
David Coyle
Charles O. Crain
Lawrence Frank Crowder
Bobbie Allen Culbertson
Ralph Darst
Lester Dees
Herman Deloach
Charles Dorris
Elbert Dunigan
Lewis Dunigan
Guy Dunn
Roy Dunn
Charles Edwards
Helen Ensey
P. W. Eubanks
Rayford Farrington
Rayford Charles Farrington
Willard C. Farrington
Carlton Ferguson
Lavelle Ferguson
Bernice Flanakin
Marion Bradford Florey
Philip Frazier
Robert L. Frazier
Leon Galbraith
Newton Galbraith
Travis Galbraith
George Gleason
Jewel Goodman
Herschel Graham
Fred Hackler, Jr.
Eddie Hair
Harry C. Hair
R. E. Hair, Jr.
William Max Hair
Jerry Harris
J. A. Harvey
Andrew Haynes, Jr.
Joe Herschell
A. C. Higginbotham
Percel Hill
Ted Hodge
Horace B. Holden
Linwood Holden
James Holland

C. T. Holloway
Austin Juckett
Otha Juckett
William Juckett
Kenneth A. Kendrick
Ernest Hoyle Knight
Charles Krouse
Emilie Krouse
E. S. Krouse, Jr.
J. D. Krouse
Jack Krouse
Victor Krouse
Lloyd LeCroy
Gene Malone
Roy B. Malone
Liberty "Frenchie" Manuel
Gordon Martin
James R. Martin
Odel "Dink" McDonald
Winfred McDonald
Junius McGill
James McLaren
Vernon Mize
William Monroe
Dennis McMullan
C. R. McWilliams
Andrew Miller
Cecil Miller
Dudley Miller
Charles Miller
Frank Miller
L. W. Miller
Thomas Miller
Jimmie Floyd Mims
Charles Mitchell, Jr.
Vernon Mize
Richard Montgomery
James A. Morgan
Wallace Morgan
Louis Mouser
George Mouser
Russell Nations
O. J. Newberry
George Nickerson
Archie Null
Morris Parker
Preston Parker
Edward Pharr
Perry Carr Pittman
Moses Pittman
Cecil Posey
Elmer Posey
William Posey
Elmer Poter
William Milton Powell
Arley Powers
Benard Powers
Jimmy Lee Powers
Maurice F. Price
Don Reynolds
Glendon Rice
James R. Robinson Jr.
Wallace Robinson
Martin Rouse
Chester Rowland
Earlene Rowland
Robert Rowland
C. W. Scott
Abner Slack
Edwin Slack
Jeral Slack
Melvin Slack
Paul Slack
Robert Slack
Rollin Slack
Wayman Slack
William Slack
Bernice Reese Smith
Harvey Smith
Isaac Franklin Smith
James Carlton Smith
Otis Smith
Robert Charles Smith
James Spencer
Charles H. Staten
Jackson Stiles
Zolin Stiles
Conrad Taylor
Billy Twitty
James Twitty
Melton Tibbals
Warren Tibbals
Perry Timmons
Allen Tomlin
Lewis Pascal Townsend
Harold Turner
Billy Tyler
Delos Tyler, Jr.
Earl Tyler
Elvin Tyler
Benjamin Julius Vaughan
Dillon Wallace
Monroe Warlick
Hardy Eugene Waters
Milton White
Lee Jesse Wortham
Bert Wheeler
Barry White
James M. White
Ralph White
Rupert T. White
James Whipple
Thomas Whipple
Bernie Watson
Paul Wilbur
D. C. Wimberly
John Wise
William Harwell Wise
Melvin Wright
Leland Wynne
Richard Wynne
Howard Zimmerman
W. F. Zimmerman

Notes

AUTHOR'S PREFACE

1. James Dalton Morrison, ed. *Masterpieces of Religious Verse* (New York,Harper and Row, 1948) 22.

2. Kate Butler, *Story of Springhill, Louisiana* (Springhill: Unpublished Manuscript, 1948) 5–6.

3. Mrs. Prescott Krouse, *Respect for the Past, Confidence in the Future: Webster Parish Centennial, 1871–1971,* Minden: *Minden Press,* 1971.

CHAPTER 1

1. Charles Robert Goins and John Michael Caldwell, *Historical Atlas of Louisiana* (Norman: University of Oklahoma Press, 1995) 9.

2. *Ibid.*, 9.

3. *Ibid.*, 3.

4. *Ibid.*, 5.

5. Phillip Cook, "The North Louisiana Upland Frontier: The First Three Decades," *North Louisiana Volume One To 1865: Essays on the Region and Its History,* B. H. Gilley, ed. (Ruston: McGinty Trust Fund Publications, 1984), 24.

6. *Ibid.*, 24–25.

7. *Ibid.*, 24–25.

8. *Ibid.*, 43.

9. *Ibid.*, 44–45.

10. *Ibid.*, 45.

11. *Ibid.*, 45.

12. Isaac Murrell, *Webster Tribune,* March, 1879.

13. Anna Cannaday Burns, *The Kisatchie Story: A History of Louisiana's Only National Forest* (Ann Arbor: University Microfilms International, 1982) 1.

CHAPTER 2

1. Robert Newman and Nancy W. Hawkins, *Louisiana Pre-History,* 2nd ed. (Baton Rouge: Division of Archeology, Department of Culture, Recreation and Tourism, 1990) 1.

2. Fred Kniffen and Sam Bowers Hilliard, *Louisiana: Its Land and Its People* (Baton Rouge: Louisiana State University Press, 1988) 104.

3. *Ibid.*, 22–25.

4. Clarence Webb and Hiram F. Gregory, *The Caddo Indians of Louisiana,* 1214.

5. Charles Robert Goins and John Michael Caldwell, *Historical Atlas of Louisiana* (Norman: University of Oklahoma Press, 1995) 23.

6. *Ibid.*, 26.

7. Margaret Ramsey Wilder, *History of Union Parish 1865–1900* (Unpublished M.A. Thesis, Louisiana Tech University, 1971) 1.

8. *Ruston Daily Leader,* September 26, 1973.

9. Ethelle and Baker Colvin, *Colvin and Allied Families* (El Dorado, Arkansas, 1965) 316.

10. Philip Cook, "North Louisiana Upland Frontier," *North Louisiana Vol. One To 1865*, ed. B. H. Gilley (Ruston: McGinty Trust Fund Publications, 1984) 29.

11. Kniffen and Hilliard, *Louisiana: Its Land and Its People,* 121.

CHAPTER 3

1. Tommie Blunt O'Bier. *North Webster Parish: The Early Years* (Shreveport: Insty-Prints, 1996) 9.

2. Archer H. Mayor, S*outhern Timberman: The Legacy of William Buchanan* (Athens: University of Georgia Press, 1988) 13–14.

3. Tommie O'Bier. *North Webster Parish: The Early Years* (Shreveport: Insty-Prints, 1996) 5.

4. *Ibid.*, 4.

5. *Ibid.*, 9.

6. Mrs. Prescott Krouse, *Respect for the Past, Confidence in the Future: Webster Parish Centennial 1871–1971,* (Minden: *Minden Press,* 1971) 42.

7. Susan Herring, ed., "Father of Claiborne Parish," *The Guardian Journal,* April 29, 1999.

CHAPTER 4

1. "Pine Woods Lumber Company Charter," Lafayette County, Arkansas, February 1897.

2. Archer H. Mayor, *Southern Timberman: The Legacy of William Buchanan* (Athens: University of Georgia Press, 1988) 38.

3. Kate Butler, *Story of Springhill, Louisiana,* 1948 (Springhill: Unpublished Manuscript, 1948) 2.

4. Tommie O'Bier, *North Webster Parish: The Early Years* (Shreveport: Insty-Prints, 1996) 12.

5. Robert Bryan, interview, 9 August 2000, Springhill, Louisiana.

6. Mark Anthony, conversation, 7 November 2000, Springhill, Louisiana.

7. Harry Brewton, ed., *The Springhill Digester* (Springhill: International Paper Co., Spring, 1979) 4.

8. *Ibid.*, 5–7.

CHAPTER 5

1. "Let's Look At Springhill," published brochure, June 11, 1948, 2.

2. *Springhill Press & News Journal,* June 11, 1948, 1–3.

3. *Ibid.*, 1.

4. *Ibid.*, 1.

5. Harry Brewton, ed. *The Springhill Digester,* 4.

6. *Springhill Press & News Journal,* October, 1978.

7. *Digester Ibid.*

8. *Ibid.*, 1.

9. Page Williamson, interview, 27 October 2000, Memphis, Tennessee.

10. *The Digester,* Fall, 1978, 4.

CHAPTER 6

1. Louisiana Municipal Review, November, 1960, 14.

2. *Prosperity from the Piney Woods,* Kansas City Southern Lines News folder, 3–4.

3. *Louisiana Municipal Review,* August, 1963, 22.

4. Tommie Blunt O'Bier, *North Webster Parish: The Early Years* (Shreveport: Insty—Prints, 1996), 13.

5. Kate Butler, *Story of Springhill, Louisiana,* Springhill: Unpublished Manuscript, 1948).

6. Charles McConnell, interview, 27 October 2000, Springhill, Louisiana.

7. *Newsfolder,* July 1952, 8.

8. *Community Progress,* 1944, 41–6.

9. Darwin Spearing, *Roadside Geology of Louisiana* (Missoula, Montana: Mountain Press Publishing Company, 1995), 25.

10. J. Fair Hardin, *Northwestern Louisiana: A History of the Watershed of the Red River 1914–1937* (Shreveport: The Historical Record Association, 1965), 404.

11. *Ibid.*, 407.

12. *Ibid.*, 408.

13. *Ibid.*, 409.

14. Dewey Williams, interview, 30 October 2000, Bossier City, Louisiana.

CHAPTER 7

1. George Peabody College for Teachers, *Digest of Survey Report* (Nashville: Division of Survey and Field Services, 1968) 3.

2. *Ibid.*, 28–45.

3. *Ibid.*, 4–5.

4. Richard Noles, *Webster Parish School Board Operating Budget for the Year Ending June 30, 2001* (Minden: Webster Parish School Board) 1–5.

5. T. H. Harris, *The Story of Public Education In Louisiana* (New Orleans: Printing Department of Delgado Trades School, 1924) 114.

6. Ardis Cawthon, Mrs. J. D. Brown, and Yvonne Rogers, *The History of Webster Parish Schools* (Minden: Webster Parish School Board) 24.

7. O'Bier, *North Webster Parish: The Early Years,* 142.

8. Wayne Williams, Jr., *A Short History of Webster Parish Schools* (Ruston: Louisiana Tech University, 1971) 42.

9. Richard Noles, interview, 21 August 2000.

10. O'Bier, *North Webster Parish: The Early Years,* 145.

CHAPTER 8

1. Louisiana, *The Lawrason Act.* State of Louisiana (1997 Revised) 3–1.

2. *Ibid.*, 3–6.

3. *Ibid.*, 3–12 - 3–14.

4. *American Heritage Dictionary,* third ed. (1992) s.v. "Ordinance."

5. Board of Aldermen. *Code of Ordinances City of Springhill, La.* (Tallahassee, Florida: Municipal Books Corporation, 1995) xiv –xvii.

6. David Camp, interview, 23 January 2001, Springhill, Louisiana.

7. Pete Porter, interview, 22 January 2001, Springhill, Louisiana.

8. Sherrell Smith, interview, 23 January 2001, Springhill, Louisiana.

9. Mike Shultz Griffin, interview, 18 January 2001, Springhill, Louisiana.

10. Charles McConnell, interview, 03 November 2000, Springhill, Louisiana.

11. *Minden Press and News Journal.* "Community Voice" feature, 1999.

12. Police Jury Association of Louisiana. *Parish Government Structure* (Baton Rouge: Police Jury Association of Louisiana, 1999) 1.

CHAPTER 9

1. *Encyclopedia Britannica,* Vol. 11 s.v. "History of Medicine," 832–841.

2. *Journal of North Louisiana Historical Association,* Vol. XVII, No. 2–3, Spring–Summer 1986. "The History of Medicine in North Louisiana," 47.

3. *Ibid.*, 52.

4. *Ibid.*, 53–54.

5. *Ibid.*, 49.

6. Tommie O'Bier, *North Webster Parish: The Early Years* (Shreveport: Insty-Prints, 1996) 335.

7. *Ibid.*, 36.

8. *Ibid.*, 350.

9. *Ibid.*, 335.

10. Mage Sims, interview, 13 November 2000, Springhill, Louisiana.

11. Tom Craig, interview, 15 November 2000, Springhill, Louisiana.

12. *Doctor's Clinic of Springhill,* September 1999, 1.

13. Greg Simmons, interview, November 20, 2000, Springhill, Louisiana.

14. *Springhill Medical Center Completes Purchase,* Springhill Press, November 23, 2000.

15. *Ibid.*

16. Sheila Franklin, interview, 22 November 2000, Springhill, Louisiana.

17. Dr. Matthew Lane, interview, 23 November 2000, Springhill, Louisiana.

18. Dr. William Tanner, interview, 21 November 2000, Springhill, Louisiana.

19. Olin Mills, interview, 21 November 2000, Springhill, Louisiana

20. *Ibid.*

21. Dr. William Odom, interview, 12 December 2000, Springhill, Louisiana.

CHAPTER 10

1. Peter Drucker. *The Effective Executive* (New York: Harper Colophon Books, 1966) viii.

2. Cindy Hall, interview, 06 March 2001, Springhill, Louisiana.

3. Margaret Edens, Interview, 03 March, 2001, Springhill, Louisiana.

4. Eluida Flanakin, interview, 01 March 2001, Springhill, Louisiana.

5. Joe Curtis, interview, 27 February 2001, Springhill, Louisiana.

6. Dr. Charles Payne, interview, 06 March 2001, Springhill, Louisiana.

7. Dr. A. C. Higginbotham, interview, 27 February 2001, Springhill, Louisiana.

8. Johnnie Hill, interview, 26, February 2001, Springhill, Louisiana.

9. Bob Colvin, interview, 27 February, 2001, Springhill, Louisiana.

10. *Springhill Press,* 22 February 2001, 6.

11. Betty Brown, interview, 02 March 2001, Springhill, Louisiana.

12. Dathene Brown, interview, 02 March 2001, Springhill, Louisiana.

CHAPTER 11

1. *The American Heritage Dictionary of the English Language,* Ann Soukhanov, ed. (New York: Houghton Mifflin Company, 1992)

2. Calvin Craig, interview, 01 March 2001, Springhill, Louisiana.

3. Mac Pace, interview, 01 March 2001, Springhill, Louisiana.

4. Ophelia Carroll, interview, 12 March 2001, Springhill, Louisiana.

5. Jack Montgomery, speech at Springhill High School Football Banquet, 22 January 2001.

6. Harry Brewton, ed., *The Springhill Digester* (Springhill: International Paper Company's White Paper Group, Spring 1979) 4.

7. Adene Mouser, interview, 13 March 2001, Springhill, Louisiana.

8. Phillip Cook, "The North Louisiana Upland Frontier: The First Three Decades," *North Louisiana Volume One to 1865: Essays On the Region and Its History,* B. H. Gilley, ed. (Ruston: McGinty Trust Fund Publications, 1984) 24–25.

CHAPTER 12

1. William Poe, "Religion and Education in North Louisiana, 1800–1865," *North Louisiana Volume One: to 1865: Essays On the Region and Its History,* B. H. Gilley, ed. (Ruston: McGinty Trust Fund Publications, 1984) 115.

2. *Ibid.*, 114.

3. *Ibid.*, 117.

4. Published and unpublished histories of all churches were used for resource material to write the individual church histories in this chapter.

5. This information was supplied by Rev. Kenneth Everett, Pastor Emeritus of Central Baptist Church.

CHAPTER 13

1. Archer H. Mayor, *Southern Timberman: The Legacy of William Buchanan* (Athens: University of Georgia Press, 1988) 166.

2. *Ibid.*, 7–8.

3. *Ibid.*, 24–25.

4. Kate Butler, *Story of Springhill, Louisiana,* (Springhill: Unpublished Manuscript, 1948) 2.
5. *Ibid.*, 3.
6. Harry Brewton, ed., *The Springhill Digester* (Springhill: International Paper Company's White Paper Group, Spring 1979) 1.
7. James Branch, interview, 04 February 2001, Springhill, Louisiana .
8. Louis Rukeyser, *What's Ahead for the Economy?* (New York: Simon and Schuster, 1983) 7.

CHAPTER 14

1. John Gardner, *Self Renewal* (New York: Harper Colophon Books, 1973) 5–7.
2. Charles Strong, interview, 13 November 2001, Shreveport, Louisiana.
3. Eddie Hammontree, interview, 21 September 2000, Minden, Louisiana.
4. Chamber of Commerce Web Site, Springhill, Louisiana (Springhill: Springhill Net, 10 May 2000) available from www.springhilllouisiana.com.
5. David Camp, interview, 23 January 2001, Springhill, Louisiana.
6. Clary Anthony, interview, 16 March 2001, Springhill, Louisiana.
7. Lumberjack Festival Association, *Lumberjack Festival Program October 9–10, 1987,* Springhill, Louisiana, 3–4.
8. Jan Willis, interview, 09 January 2001, Springhill, Louisiana..
9. *Ibid.*
10. Dorothy Smith, "International Paper Company Closes," *Springhill Press and News Journal,* October, 1978.

CHAPTER 15

1. Archer H. Mayor, *Southern Timberman: The Legacy of William Buchanan* (Athens: University of Georgia Press, 1988), 1.
2. Edwin Adams Davis, Raleigh Suarez, and Joe Gray Taylor, *Louisiana—The Pelican State* (Baton Rouge: Louisiana State University Press, 1985), 209.
3. Mayor, *Southern Timberman: The Legacy of William Buchanan,* 7–8.
4. *Ibid.*, 24 –25.
5. Richard Marshall, ed., *Great Events of the 20th Century* (New York: Readers Digest Association, Inc., 1977) 222–223.
6. Kate Butler, *Story of Springhill, Louisiana,* (Springhill, Unpublished Manuscript, 1948) 2.
7. *Ibid*, 3.
8. *Ibid.*, 3.
9. Marshall, *Great Events of the 20th Century,* 100.
10. Henry Banford Parkes, *The United States of America: A History* (New York: Alfred Knopf, 1953), 567.
11. *Springhill Press and News Journal,* May, 2000.
12. William T. "Toxie" Bowen, interview, October, 2000, Springhill, Louisiana.
13. *Springhill Press and News Journal,* September, 2000.
14. D.C. Wimberly, interview, 18 September 2000, Springhill, Louisiana.
15. *Ibid.*
16. *Ibid.*
17. James Dalton Morrison, *Masterpieces of Religious Verse* (New York: Harper and Row, Publishers, 1948) 539.

CHAPTER 16

1. Richard Marshall, ed. *Great Events of the Twentieth Century* (New York: Readers Digest Association, 1977).
2. John McWhorter, *Losing the Race—Self Sabotage In Black America* (New York: The Free Press, 2000) xi–xiii.
3. *Ibid.*
4. *Ibid.*, xi–xiii.
5. Edwin Davis, Raleigh Suarez and Joe Gray Taylor, *Louisiana—The Pelican State* (Baton Rouge: Louisiana State University Press, 1985) 39.
6. *The Britannica Guide To Black History* (Chicago: University of Chicago Press, 2000) 6.
7. *Ibid.*, 13.
8. *Ibid.*, 30.

CHAPTER 17

1. Mrs. Prescott Krouse, *Webster Parish Centennial 1871–1971,* (Minden:Minden Press, 1971) 19.
2. *Ibid.*, 20.
3. *Ibid.*, 20.
4. *Ibid.*, 20.
5. *Ibid.*, 20.
6. *Minden Press–Herald,* Februry 24, 2000, 6E.
7. *Louisiana Municipal Review,* August 1963, 23.
8. *Minden Press–Herald,* February 24, 2000, 6E.
9. *Ibid.*, 4E.
10. *Ibid.*, 3E.
11. *Ibid.*, 4C
12. *Ibid.*, 4C.
13, *Ibid.*, 4C.
14. Tommie O'Bier, *North Webster Parish: The Early Years* (Shreveport: Insty-Prints, 1996) 125.
15. *Ibid.*, 125.
16. *International Standard Bible Encyclopedia,* James Orr, ed. (Grand Rapids: William B. Eerdman's Publishing Company, 1960), 3132.
17. Emily Miller, *The History of Sarepta* (Shreveport: Toby Printing Company, 1924) 6.

18. *Minden Press-Herald,* February 24, 2000, 3C–4C.

19. *Shongaloo High School Birthday Calendar and History,* 1994, 4.

20. *Ibid.*, 4.

21. Mrs. Prescott Krouse, *Webster Parish Centennial 1871–1971* (Minden: *Minden Press,* 1971).

22. *Ibid.*

23. Mount Paran Baptist Church, Brochure, July 14, 1996.

24. Mrs. Prescott Krouse, *Webster Parish Centennial 1871–1971* (Minden: *Minden Press,* 1971).

25. *Springhill Press and News Journal,* March 31, 1976, 3C.

26. *Minden Press–Herald,* February 24, 2000, 6B.

CHAPTER 18

1. *The American Heritage Dictionary of the English Language,* 3rd ed. (New York: Houghton Mifflin Company, 1992) Ann Soukhanov, editor.

2. *Ibid.*

3. Mike Shultz Griffin, interview, 14 January 2001, Springhill, Louisiana.

4. Eugene Waters, interview, 04 August 2000, Springhill, Louisiana.

5. *Ibid.*

6. *Ibid.*

7. Olin Mills, interview, 07 February 2001, Springhill, Louisiana.

8. Tommie O'Bier, *North Webster Parish: The Early Years* (Shreveport: Insty-Prints, 1996) 13.

9. James Branch, interview, 03 February 2001, Springhill, Louisiana.

10. O'Bier, *North Webster Parish: The Early Years.*

11. *Ibid.*

12. *Ibid.*

13. James Branch, interview, 03 February 2001, Springhill, Louisiana.

14. *Ibid.*

15. *Ibid.*

16. *Ibid.*

17. Johnnie Hill, interview, 31 January 2001, Springhill, Louisiana.

18. O'Bier, *North Webster Parish: The Early Years,* 11.

19. *Ibid.*

20. *Ibid.*

21. *Ibid.*

22. Johnnie Hill, interview, 31 January 2001, Springhill, Louisiana.

23. O'Bier, *North Webster Parish: The Early Years.*

24. Story told by Jerry Bonner to author when they were young boys, Springhill, Louisiana.

25. Olin Mills, interview, 21 November 2000, Springhill, Louisiana.

26. Scott Boucher, interview, March 2001, Springhill, Louisiana.

27. Evelyn McMullan Colvin, interview, March 2001, Minden, Louisiana.

28. *Ibid.*

29. Dennis McMullan, interview, March 2001, Springhill, Louisiana.

30. *Ibid.*

31. Mickey Parker, interview, March 2001, Jackson, Mississippi.

32. Dennis McMullan, interview, March 2001, Springhill, Louisiana.

33. Evelyn McMullan Colvin, interview, March 2001, Minden, Louisiana.

34. Author's research of church histories.

35. Author's research of early medicine in Springhill. Louisiana.

CHAPTER 20

1. Will Durant, *The History of Civilization,* Vol. 4: *The Age of Faith* (New York: Simon and Schuster, Inc., 1950) 979.

CHAPTER 21

1 Will Durant, *The Lessons of History* (New York: Simon and Schuster, 1968) 91.

2. Frances Page Williamson, "Graphic Memories of the City's History," *Springhill Press and News Journal,* November 4, 1987, 12B.

3. Robert Frost, "Stopping by Woods on a Snowy Evening," *Complete Poems by Robert Frost* (New York: Holt, Rinehart, and Winston, 1949) 275.

APPENDIX J

1. American Legion, *Men and Women in the Armed Forces during World War II from Webster Parish* and information from friends and family. Every effort was made to obtain an accurate listing. Any omission is regretted.

Sources Consulted

BOOKS

Angelou, Maya. I *Know Why the Caged Bird Sings.* New York: Bantam Books, 1980.

Brinton, Crane, John B. Christopher, and Robert Lee Wolff. *A History of Civilization,* Vol. 1. Englewood Cliffs: Prentice-Hall, 1955.

Brown, Claira. *Commercial Trees of Louisiana.* Baton Rouge: Louisiana Office of Forestry, 1972.

Burns, Anna Cannaday. *A History of Kisatchie National Forest.* Pineville: Offices of the Kisatchie National Forest, 1981.

Burns, Anna Cannaday. *A History of the Louisiana Forestry Commission.* Natchitoches: Louisiana Studies Institute, Northwestern State College, 1968.

Burns, Anna Cannaday. *Fullerton: The Mill, The Town, The People, 1907–1927.* Alexandria: The author, 1970.

Burns, Anna Cannaday. *The Kisatchie Story: A History of Louisiana's Only National Forest.* Ann Arbor: University Microfilms International, 1982.

Carson, Rachel. *Silent Spring.* Boston: Houghton Miffin, 1962.

Carter, Cecile Elkins. *Caddo Indians: Where We Come From.* Norman: University of Oklahoma Press, 1995.

County Clerk of Webster Parish. *Charters, Bank of Springhill, November 4, 1916.* Minden: County Clerk's Office, 1916.

County Clerk of Webster Parish. *Miscellaneous Acts, Book B.* Minden: County Clerk's Office, September 25, 1922.

Davis, Edwin Adams, Raleigh Suarez, and Joe Gray Taylor. *Louisiana—The Pelican State.* Baton Rouge: Louisiana State University Press, 1985.

Division of Surveys and Field Services. *Webster Parish Public Schools.* Nashville: George Peabody College for Teachers, 1969.

Durant, Will and Ariel Durant. *The Lessons of History.* New York: Simon and Schuster, 1968.

Franks, Kenny and Paul Lambert. *Early Louisiana and Arkansas Oil—A Photographic History.* College Station: Texas A&M. University Press, 1946.

Frost, Robert. *Complete Poems of Robert Frost.* New York: Holt, Rinehart and Winston, 1964.

Gibson, Joe L. *Poverty Point.* Baker, Louisiana: Department of Culture, Recreation and Tourism of Louisiana, 1996.

Gilley, B.H., ed. *North Louisiana Volume One to 1865: Essays on The Region and Its History.* Ruston: McGinty Trust Fund Publications, 1984.

Goins, Charles Robert and John Michael Caldwell. *Historical Atlas of Louisiana.* Norman: University of Oklahoma Press, 1995.

Great Books of the Western World. "Biographical Note Thucydides," Vol. 6. Chicago: Encyclopedia Britannica, Inc., 1952.

Hardin, J. Fair. *A History of the Watershed of the Red River 1714–1939.* Vol. 1. Baton Rouge: Louisiana State University Press, 1939.

Harris, D. W. and B. M. Hulse. *History of Claiborne Parish, Louisiana.* New Orleans: Press of W. B. Stansbury and Company, 1886.

Harris, T. H. *The Story of Public Education In Louisiana.* New Orleans: Printing Department of Delgado Trades School, 1924.

Kniffen, Fred and Sam Bowers Hilliard. *Louisiana:*

Its Land and People. Baton Rouge: Louisiana State University Press, 1988.

Kniffen, Fred and Hiram F. Gregory. *The Historic Indian Tribes of Louisiana: From 1542 to the Present.* Baton Rouge: Louisiana State University Press, 1987.

Krouse, Mrs. Prescott. *Webster Parish Centennial 1871–1971.* Minden: Minden Press, 1971.

Limerick, Patricia Nelson. *The Legacy of Conquest.* New York: Norton and Company, 1987.

Mayor, Archer H. Southern Timberman: *The Legacy of William Buchanan*. Athens: University of Georgia Press, 1988.

McGinty, Garnie William. *A History of Louisiana.* New York: The Exposition Press, 1949.

McGinty, Garnie William. *A Twig of the McGinty Family Tree.* Bossier City: Tri-State Printing and Bindery, 1979.

Miller, Emily V. *The History of Sarepta.* Shreveport: Toby Printing Company, 1924.

Newman, Robert and Nancy W. Hawkins. *Louisiana Pre-History,* 2nd ed. Baton Rouge: Division of Archeology, Department of Culture, Recreation and Tourism, 1990.

Noles, Richard. *Webster Parish School Board Operating Budget for the Year Ending June 30, 2001.* Minden: Webster Parish School Board, 2000.

North Webster Chamber of Commerce. *Welcome to Springhill.* Springhill: Springhill Press, 1997.

O'Bier, Tommie Blunt. *North Webster Parish: The Early Years.* Shreveport: Insty-Prints, 1996.

Oxford Dictionary of Quotations, 2nd ed. New York: Oxford University Press, 1955.

Paxton, W.E. *A History of Baptists In Louisiana, From Earliest Times To the Present* (St. Louis: C. R. Barnes Publishing Company, 1888).

Paret, Ory G. *History of Land Titles In State of Louisiana*. U.S. Statutes Vol. 2, p. 283; Vol. 9, p. 123; Vol. 12, p. 392.

Polk's Springhill City Directory. Dallas: R. L. Polk and Co., 1964.

Statistical Abstract of Louisiana 1990, 8th ed. New Orleans: University of New Orleans, 1990.

Statistical Abstract of United States 1997. Washington: U.S. Department of Commerce, 1997.

Succession of J. F. Giles. Minden: Webster Parish Clerk's Office.

Trees—Louisiana's No. 1 Crop. Alexandria: Louisiana Forestry Association, 1979.

Wall, Bennett H., ed. *Louisiana: A History.* Arlington Heights: The Forum Press, Inc., 1984.

Wahlenburg, W. G. *Longleaf Pine: Its Use, Ecology, Regeneration, Protection, Growth and Management.* Washington: Charles Lathrop Pack Forestry Foundation, 1946.

Webb, Clarence H. and Hiram F. Gregory. *The Caddo Indians of Louisiana*. Baton Rouge: Archeological Survey and Antiquities Commission, 1990.

Williams, Wayne W., Jr. *A Short History of Webster Parish Schools.* Ruston: La. Tech University, 1971.

JOURNALS

"Religion In Webster Parish," North Louisiana Historical Association *Journal*, Spring 1973, Vol. 4, No. 3, pp. 84–89.

"The History of Medicine In North Louisiana," North Louisiana Historical Association *Journal,* Spring–Summer 1986, Vol. XVII, No. 2–3.

NEWSPAPERS

Springhill Press and News Journal, June 30, 1976, pp. 6E–10E.

Springhill Press and News Journal, November 4, 1987, p. 12B

Springhill Press and News Journal, June 29, 1977, p. 1.

Springhill Press and News Journal, October 18, 1978, pp. 1–5.

Springhill Press and News Journal, March 14, 1979, p. 2C.

Frances Page Williamson, "Graphic Memories Recount City's History," *Springhill Press and News Journal,* November 4, 1987, p. 12B.

Springhill Press and News Journal, May 8, 2000.

Springhill Press and News Journal, June 15, 2000.

Springhill Press and News Journal, August 17, 2000.

UNPUBLISHED MATERIALS

Butler, Kate, "Story of Springhill, Louisiana," 1948.

PAMPHLETS

Chamber of Commerce, *Welcome To Springhill,* September, 1977.

INTERVIEWS

Anthony, Clary. 16 March 2001, Springhill, Louisiana.

Bailey, Pat. 14 June 2001, Springhill, Louisiana.

Bankhead, Edward. 15 January 2001, Springhill, Louisiana.

Boucher, Jesse. 15 February 2001, Springhill, Louisiana.

Boucher, Scott. March 2001, Springhill, Louisiana.

Bowen, Toxie. 02 October 2000, Springhill, Louisiana.

Branch, James. 03 February 2001, Springhill, Louisiana.

Branton, Georgia. 09 November 2000, Shongaloo, Louisiana.

Brown, Dathene. 02 March 2001, Springhill, Louisiana.

Camp, David. 23 January 2001, Springhill, Louisiana.

Carroll, Ophelia. 12 March 2001, Springhill, Louisiana.

Colvin, Bob. 27 February 2001, Springhill, Louisiana.

Colvin, James. 16 March 2001, Springhill, Louisiana.

Craig, Calvin. 01 March 2001, Springhill, Louisiana.

Craig, Thomas. 15 November 2000, Springhill, Louisiana.

Curtis, Joe. 01 September 2000, Springhill, Louisiana.

Edens, Margaret. 03 March 2001, Springhill, Louisiana.

Flanakin, Eluida. 27 September 2000, Springhill, Louisiana.

Franklin, Sheila. 22 November 2000, Springhill, Louisiana.

Garrison, Avis. 22 February 2001, Springhill, Louisiana.

Griffin, Mike Shultz. 14 January 2001, Springhill, Louisiana.

Hall, Cindy. 06 March 2001, Springhill, Louisiana.

Hammontree, Eddie. 21 September 2000, Minden, Louisiana.

Hardy, Evelyn. 05 October 2000, Taylor, Arkansas.

Herrington, Johnny. 14 September 2000, Springhill, Louisiana.

Higginbotham, A. C. 06 April 2001, Springhill, Louisiana.

Hill, Johnnie. 31 January 2001, Springhill, Louisiana.

Huddleston, Ray. 05 April 2001, Springhill, Louisiana.

Jackson, Charles. 23 September 2000, Springhill, Louisiana.

King, Basil. 04 October 2000, Cullen, Louisiana.

Lane, Matthew. 23 November 2000, Springhill, Louisiana.

McConnell, Charles. 27 October 2000, Springhill, Louisiana.

McDonald, Rachel. 06 October 2000, Springhill, Louisiana.

McEachern, Mary. 27 September 2000, Springhill, Louisiana.

McMullan, Dennis. 06 April 2001, Springhill, Louisiana.

Mills, Olin. 21 November 2000, Springhill, Louisiana.

Mouser, Adene. 13 March 2001, Springhill, Louisiana.

Noles, Richard. 21 August 2000, Minden, Louisiana.

Odom, William. 12 December 2000, Springhill, Louisiana.

Pace, Mac. 01 March, 2001, Springhill, Louisiana.

Payne, Charles. 06 March 2001, Springhill, Louisiana.

Porter, Peter. 22 January 2001, Springhill, Louisiana.

Price, Maurice. 20 September 2000, Springhill, Louisiana.

Reynolds, Don. 05 April 2001, Springhill, Louisiana.

Rhone, Henry. 06 March 2001, Springhill, Louisiana.

Simmons, Greg. 20 November 2000, Springhill, Louisiana.

Sims, Mage. 13 November 2000, Springhill, Louisiana.

Smith, Sherrell. 23 January 2001, Springhill, Louisiana.

Soileau, Marvin. 15 June 2001, Springhill, Louisiana.

Strong, Charles. 15 January 2001, Shreveport, Louisiana.

Tanner, William. 21 November 2000, Springhill, Louisiana.

Thomas, Jimmy. 15 January 2001, Springhill, Louisiana.

Washington, Jonathan. 09 November 2000, Springhill, Louisiana.

Waters, Eugene. 04 August 2000, Springhill, Louisiana.

Williams, Dewey. 30 October 2000, Benton, Louisiana.

Williamson, Page. 27 October 2000, Memphis, Tennessee.

Willis, Jan. 09 January 2001, Springhill, Louisisna.

Wilson, Wilbur. 26 February 2001, Springhill, Louisiana.

Wimberly, D. C. 18 September 2000 and 14 June 2001, Springhill, Louisiana.

Index

Adkins, Jesse Mae, 201
African-American culture, 186
Ahmed, Dr. Mahamed, 98
Alden, Isaac, 14
Allen, Clory Haynes, 92
Allen, Connie, 112
Allen, Grover C., 198
Allen, J. I., 198
Allen, James, 162
Allen, Mrs. A. J., 197
Allen, Mrs. Leland, 162
Alley, Barbara, 133
Alvis, Max, 121
American Legion, 235
American Legion Auxiliary, 227
American Legion baseball, 120
American Tri-State Underwriters, 173
Andrews, Irma, 29
Andrews, Robert, 140
Anthony-Bryan Insurance Company, 24, 220, 238
Anthony, Clary, 23, 26, 53, 219, 220
Anthony Forest Lumber Company, 163
Anthony Forest Products, 23, 160, 220
Anthony, Frank, 23
Anthony, Melvin, 23, 26, 53
Arizona Chemical Company, 50, 56
Ark-La Partners in Genealogy, 110
Arnold McLaren's barber shop, 26
Atkins, Bachmon, 199
Atkins, Trace, 198
Avalon Technology, 162
Axe Café, 216

B&S Supply, 223
Bailey, Carrol, 136
Bailey, Cecil, 42, 43
Bailey, Esther Pat, 220
Bailey Mortuary, 171
Bailey, R. A., 220
Bailey, Rose, 111
Baker, John, 215, 218
Baker, Richard, 78
Bank of Sarepta, 198
Bank of Springhill, 25, 26, 27, 52, 158
Bankhead, Ed, 78, 189, 221
Baptists, 135, 136, 138, 201, 217
Barefoot, Louisiana, xiii, 9, 16, 17, 26, 224, 243
Barnard, Bob, 29
Barnard, Lewis, 115
Barnard, Riley, 29
Barnes, Robert, 182
Basham, Eugene, 180
Baucum, William "Billy," 122, 123, 130, 131, 132, 221
Baucum, Delia Lizabeth Slack, 221
Baucum, Oscar Fulton, 221
Baucum, W. D., 68
Be Bops Grocery and Market, 195, 196
Beaty, Howard, x, 52
Beaty, Jr., Howard, 198
Beauchamp, Rev. Frank, 140
Beautification Committee, 111, 238
Benefield, B. J., 115
Benson, Dr. Kathryn M., x
Berry, L. E., 211
Black Mason Hall, 69
Blackwell, Vivian, 115
Blanton, Tylon, 41, 83, 97, 222
Blocker, Rev. R. M., 141
Blocker's Chapel, 67

Blount, Tom 180
Bodcaw Lumber Company, 16, 18, 24, 51, 137, 163, 197, 224, 243
Bolden, Elder Burnice, 143
Bond Drug Store, 101
Bonner, Andrea, iii, xi
Bonner, Ann, iii, x, 111
Bonner, Ann H., xi
Bonner, Christopher, iii
Bonner, Curtis, iii
Bonner, Flossie, iii, 23
Bonner, Flossie Tapscott, 29
Bonner, Gary, xi, xiii, 132
Bonner, Jerry, 215
Bonnie and Clyde, 216
Bootleg whiskey, 216
Bordelon, Rev. Vernon, 139
Boucher and Slack Insurance, 53, 60
Boucher Drug, 60, 62, 102
Boucher, Erma, 115
Boucher, Gary, 68
Boucher, Gus, 68
Boucher, Jesse, 26, 68, 102, 164, 222, 223, 228
Boucher-Lytle, Sherry, 68
Boucher, Melvin, 215, 218
Boucher, State Senator Drayton, 104
Bowen, William T., 180
Bowles, Robert, 121
Boy Scout Hut, 87
Boyer, A. T., 140
Boyette, Burton B., 182
Bradley, Jr., Superintendent Stephen, 143
Brady, J. E., 140
Brady, John, 199
Brady, Mrs. John, 139
Branch, Annie Lee, 223
Branch Brothers Ford Motor Company, 53, 212, 213, 223
Branch, J. A., 52, 211, 212
Branch, J. B., 211, 212
Branch, James, 26, 54, 213
Branch, Mrs. J. C., 162
Branch, Sr., J. B., 223
Branton, Georgia, x, 200
Branton, Parey, 200, 201
Breaux, Carroll, 78
Brewer, Elizabeth, x
Brewton, Harry, 39
Bridwell, J. C., 199
Bridwell, Jake, 201
Bridwell, Verna, 201
Brookshire's, 166
Brossette, Rev. Jimmy, 142
Brown, Charles Herbert, 69, 191
Brown, Clayton Cornish "Cracker" Brown, 122
Brown, E. W., 45, 97, 122
Brown High School, 69, 188
Brown, J. D., 115
Brown, J. W., 195
Brown Junior High School, 188
Brown, Mary, 142
Brown Middle School, 237
Brown, Rev. Harrison, 138
Brown, Rev. T. H., 138
Brown, Sandra, 196
Brown, Tommy, 142
Browning, Dr. Joe, 224
Browning, Dr. Melvin, 224
Browning Elementary School, 76, 222, 239
Browning, J. M., 52
Browning, John, 26, 30, 224
Browning, John Marvin, 224
Browning, Joseph Robert, 96
Browning, M. L., 113
Browning, Merit Taylor "Mitt," 52, 53, 67, 76, 224
Bryan, Billy, 24
Bryan Insurance Agency, 24
Bryan, Barbara Barnard, xi
Bryan, Rex, 52
Bryan, Robert C., xi
Bryan's Shoe Store, 53
Bryant, Bear, 132
Buchanan, J. A., 16
Buchanan, William, 16, 19, 21, 69, 157, 178, 197, 212, 213, 224, 243
Budwah, Rev. Carl, 137
Burke, Syvell, 215
Burnham, M. C. "Doc," 102, 195
Burnham's Drug, 102
Burns, Dr. Purnel, 230
Burns, Mildred, 201
Burt, Jasper, 195
Burton, A. M., 136
Bush, Bob, 99
Butler, Dr. Rupert, 26, 52, 214, 217
Butler, Flo, 217
Butler, James, 197
Butler, Kate, ix, 23, 104, 158, 178
Butler Memorial, 106
Butler Memorial Health Center, 107, 225
Butler Memorial Health Unit, 233
Butler, Rupert, 15, 21, 22, 52, 96, 104, 178, 225
Byrd, Jack, 215
Byrnes, J. A., 15

C. C. Harper's Texaco Consignee, 53
C.&S. Ready Mix, 50
Caddo Indians, 6, 7, 199
Caddo medicine men, 95
Cadenhead, L. M. "Cotton," 100, 215
Calvary Baptist Church, 195
Camelot, 189
Camp, David, 80
Campbell, Joe, 215

Capers, Candalie La Mourne, 69
Caraway, Helen, 111
Carmack, Jr., James H., 182
Carroll, Ophelia, 68, 122
Carter, Cliff, 113
Carter, Jerry, 140
Carter, Mrs. Sarepta, 15, 197
Cason, Charles W., 182
Cassells, Jennings, 195
Castleberry, Carole, x, 111
Centennial Celebration Committee, 228
Central Baptist Church, 136, 144, 228, 233, 237
CenturyTel, 164
Chadwick, Jimmy, 1
Chamber of Commerce, 164, 172, 230, 231, 233, 235, 238
Charles H. Brown Middle School, 77, 191, 192
Charles H. Brown School, 71
Chase Studio, xi
Chastant, Buddy, 182
Chen, Dr. Cecilier, 98
Choate, Attorney, 52
Christmas parade, 123
Chumley, Dr. Gary, 100
Church of Christ, 197
Church of Christ on N. Arkansas Street, 148
Church of Christ on Butler Street, 146
Church of God, 195
Church of God in Christ, 195
Circle S Riding Club, 128
Citizens Bank and Trust, 52, 62, 162, 170, 220, 230
City Hall, 93
Civic Center, 93, 201
Civic Club, 111
Civilian Conservation Corps, 231
Clabaugh, —, 122
Claiborne Parish, 199
Clark Burnes Downtown Store, 200
Clements, Ben, 123
Clements, E. L., 140
Clifford, Louisiana, 38, 194, 195, 213, 226
Coleman, John, 69
Collins, Rev. G. R., 137
Colored First Baptist Church, 195
Columbia Hospital, 99
Colvin family, 8
Colvin, Bob, 99
Colvin, Caldwell, x
Colvin, Floy DeLoach, 225
Colvin, Malcolm C., 225
Commercial Bank and Trust Company, 158
Commercial Bank and Trust Company Depository, 52
Community Activities Center, 80, 81
Community House, 87
Compulsory Attendance Law, 67
Cone, Kermit, 115
Cone, Retha, 115
Congress of Racial Equality, 188
Consolidated Chemical Industries of American Cyanide, 50
Cook, Dana, 111
Cook, Philip, 2
Cooper, Ellis, 68
Cotton Belt, 215
Couvillon, Rev. Francis O., 139
Cox, M. L., 198
Coyle and Company, 198
Coyle, Bud, 39
Coyle, George, 180
Coyle, Hugh Jacob, 15, 194
Coyle, Jr., Kenny, 196
Coyle, Kenny, 196
Coyle, Leonard, 68
Craig, Calvin, 121
Craig, Tom, x, 201
Crockett, E. M., 26
Crockett, Harry H., 26
Crouch, C. D., 136
Crow, Joe, 216
Crow, John David, 68, 121, 122, 132, 222, 225
Crusader Drilling Service, 50
Culberson, —, 122
Cullen Lions Club, 195
Cullen, Louisiana, 38, 194, 195, 196, 202, 208, 213, 216, 226, 232
Cullen, Richard J., 38, 195, 226
Curry, Donald, 70
Curtis Brothers Grocery, 53
Curtis, J. D., 195
Curtis, Jack, 195
Curtis, Joe, x, 82, 112
Curtis, M. A., 195
Curtis, Mac, 195
Custer, Noel, 132

Daigle, Rev. Karl, 139
Daniel, Elder S. D., 143
Darby, William, 199
Davis, Bill, 24
De Vriendat, Rev. Robert, 139
Dees, Andrew, 113
DeLoach, Dr. C. T., 198
Deloteus, Palmer W., 140
Denham, William, 197
Dennis, Barney, 56
Dennis, John, 99
Depression of 1893, 177
Diamond, Emery, 114
Dickey, S. B., 113
Dillon, Dr., 26, 52
Dinner Theatre, 111, 165
Ditto, Dr. Steve, 98
Dixie League baseball, 121

Doctor's Clinic, 98, 99, 109
Dooley, Jesse Calvin, 138
Dorcheat Acres Missionary Baptist Church, 137
Dorsey, Pam, 198
Downs, Rev. John J., 139
Doyles, Major John, 199, 200
Dreher, Fred, 195
Driskill, Brian E., xi
Driskill, Harlon, 111
Driskill, Jimmy Harlon, 56
Driskill, Nancy, 111
DuBois, W.E.B., 188
Duke, Tommy W., 242
Dumas, Gerry Colvin, 68
Dunnigan, L. A., 180
Durant, Will, 243, 245
Durham, J. P., 136

E.B. Smith and Sons, 50
E.S. Sikes Pipeline Contractor, 50
Eason, Eugene, 51, 162
Eason, F. F., 140
Eason, Francis, x
East, I. Y., 44
East Side Baptist Church, 137
East Side Missionary Baptist Church, 152
Edens, J. B., 111
Edens, Margaret, 111
Edwards, Ladelle, 133
Electric Melting Services, 169
Elkins, Rev. Dean, 142
Emancipation, 187
Emmons, Mrs. S. R., 162
Ensey, R. L., 26, 113
Ensey, Ralph, 162, 215
Entergy, 164
Ervin, Jean, x, 52
Essig, George, 212
Evening Star Methodist Church, 195
Everett, Rev. Kenneth, 136
Everett, Reverend Kenneth, x
Ex-Prisoners of War, 239

Farmer, Stanley, 199
Farmer, William, 15
Farrar, Travis, 68, 123, 134, 226
Farrell, Therral, 80
Ferguson, Carlton, 182
Ferguson, Reese, 115
Ferguson, W. T., 16
505 Service Station, 53
First Assembly of God, 137, 150
First Baptist Church, 137, 138, 147, 195
First Baptist Church of Cullen, 207
Fissel, Shorty, 215
Flanakin, Eluida H., xi, 227
Flanigan, Wilbert, 123
Flo's Café, 200
Flowers, Pompey, 21
Ford, Time, 195
Formby, Stanley, 201
Fortenberry, Rev. Luther, 136
Foster, Arthur, 24
Fountain View, 100
Fountain View Care Center, 108
Fowler, Steve, 198
Frank Anthony Memorial Park, 24, 81, 112, 117, 163, 216, 220, 238
Fred's Pharmacy, 101, 102
Fritz, Lois Turner, x, xi
Frost Lumber Company, 24, 163, 211
Frost Lumber Industries, 23, 38, 178, 234
Frost, Robert, 246

G. F. Wacker Store, 53
Garland, G. C., 97
Garland, Thomas, 99
Garnea, Dr., 97, 98
Garrett, Dr. William, 97
Garrett's Clinic, 97
Garrison, Mrs. Avis, 123
Gayle, Jim, 215
General School Act, 66
Giles, J. F., xiii, 17, 21, 26, 69, 138, 212, 213, 216, 233
Gillespie, Marie, 111
Gilley, Dr. B. H., x
Gilley, Jeanne Mack, 68
Gladney Haynes Store, 200
Gleason Crater, 201
Gleason, W. E., 200
Godley, Stella, 111
Good Sam Club, 116
Gray Clinic, 106
Gray, Dr., 98, 100
Gray, Dr. Wilson, 97
Gray's Hospital and Clinic, 97
Great Depression, 177, 178, 244
Griffin, Richard, 198
Grisham Studio, xi

H&B Drug, 101, 102, 106, 228
Habitat for Humanity, 112, 118
Hagler family, 15
Hagler, Jesse, 8
Hair, Billie, 116
Hall, Cindy, x, 111
Hall of Fame, 68, 164
Hall, Rev. M. L., 137
Hall school, 200
Halterman, A. A., 195
Hand, Dr., 98
Hanson, Joshua, xi
Hardy, Evelyn Krouse, xi

Hardy, Evelyn, x
Harper, Clyde C., 140
Harper, Lelia, iii
Harris, Dr., 215
Harris, Dr. Edward Lee, 101, 121
Harris, Elder Earl, 143
Harrison Chapel Baptist Church, 138, 150, 189, 221
Harrison, William, 15, 200
Harrison, William D., 199
Harvard University, 236
Harvey, Dr. Rhonda, 102
Harvey, Dr. Ron, 136
Haynes District, 67
Haynes, Drew, 201
Haynes, Earl, 68
Haynes, Gussie, 201
Haynes, J. T. "Sleepy," 121
Haynes, Michael, 68
Haynes, Reuben, 28
Haynes, Vickey, 112
Hearn, Craig, 201
Heintze, Dr., 103
Hemphill, Roger, 201
Henderson, Evan, 142
Herbert Owen Park, xi
Herrington, Johnny, 26
Herrington, John D., 40, 68, 78, 82, 99, 163, 214, 215, 227
Herrington, Mayor Johnny, x
Higginbotham, Dr. A. C., x, xi, 102, 105, 228
Hilburn, Frances McGowen, 68
Hilburn, J. C., 115
Hilburn, Robert, 78
Hill, Esther, 115
Hill, Johnnie, 26, 52, 214, 215
Hisaw, Tim, 139
Hodge, B. J., 68
Hodge, Rev. Jesse, 138
Hodgkiss, Dr., 101
Holiday Motel, 216
Holladay, Dr. Sam, x, 98
Holladay, Jr., Dr. Sam, 99
Holland, Andrew Bryan, 180, 183
Holland, Edward Graham Jackson, 180, 184
Holland, James J., xi
Holland, John, 80
Holland, Luther, 180
Holland, William Luther Sevier, 183
Hollingsworth, Rev. J. E., 139
Holloway, Betty, 111
Holy Temple Church of God in Christ, 195
Hood, Henry, 39
Hornbuckle, E. T., 140
Horne, Martha Machen, xi
Hornsby, Henry, 115
Hotel Manuel O'Garte, 26
Houston, L. L., 197
Howell Elementary School, 75, 234
Howell, Georgia, 75, 76
Huddleston, Ray, x, 52
Hudson, Ray, 122
Huey family, 8
Hull, Emmett, 45
Hull Furniture, 234
Hull, Jon, 142
Hull, Phil, 142
Hull, Wally, 24
Humana, 99
Humana Hospital, 98
Hunsinger, Dr. Charles, 98
Hunt, H. L., 216
Hunter, Rev. Alex W., 140

I-69 Transcontinental Highway Coalition, 222
"In Flander's Fields", 182
Indian Run Restaurant, 200
Indian Springs, 201
International Paper Company, 24, 25, 37, 38, 41, 42, 43, 120, 158, 161, 164, 180, 194, 195, 213, 226, 232, 234, 243, 244
Iron Works, 169

J. McKenzie saw gang, 20
J. N. Bond Drug Store, 53
J&M Construction Company, 236
Jackson, Charles, 52
Jackson, Geraldine, 116
Jackson, Jesse, 186
Jackson, John, 215
Jacobs, Gloria, 111
Jarrett, Tommy, 80
Jeane, David, xi, 7
Jeter, Mr., 121
Jimmy Lindsey barbershop, 200
Johnson, Avery, 215
Jones, Raymond, 142
Jones, Ruth Malone, 68

Kansas City Railroad, 51
Kauffman, Dollie, 111
Kendricks, Mrs. K. A., 142
Kennedy, R. B., 198
Kennedy, Rev. Jack M., 140
Kenyan Enterprises, 238
Kenyon, Ed., 116
Kilmer, Joyce,vi
King, Martin Luther, 186
King's Corner Assembly of God Church, 151
Kinningham, S. C., 136
Kirkpatrick, Elder A. D., 143
Knesel, Pat, x
Kottenbrook, Bill, 215
Kottenbrook, Susan, 100
Krouse, E. S., 47

Krouse, Pat, 38

L.S.U. experimental gardens, 46
L&A Railroad, 16, 51, 197, 216, 225, 234
Lacey, Ralph, 212
Lake Erling, 40, 121, 123
Lake Erling choir, 214
Lane, Dr. Matthew, 100
Lavent, Bob, 215
Law, Dr. David, 98
Leonard, Mrs. Noble, 201
Lettermen's/Letterwomen's "S" Club, 68
Lewis, Dr., 98
Lewis, Louise Wardlaw, 68, 122, 133
Life Point, 99
Lindsey, Della, 96
Lions Club, 112, 228, 230, 231, 235, 238, 239
Little Boys Baseball League, 231
Little League baseball, 121
Log Cabin Museum, 201
Long, Earl K., 86
Louisiana State Senate, 231
Love, W. M., 97
Lowes, Wilburn, 24
Loy Cox's Store, 200
Lucky Dog Café, 211
Lumberjack Café, 55
Lumberjack Festival, 123, 164, 238
Lumberjack Festival Association, 113
Lumberjack Lanes, 125
Lyles, Rev. Lewis, 139
Lynd, Butch, 123
Lynd, Willie Butch, 115
Lytle, Sherry Boucher, 228

M. T. Browning Elementary School, 224
M. T. Browning General Merchandise, 53
Mac Dunn's Café, 200
Machen, Mrs. Ella, 31
Machen, R. O. Jr., xi
Machen, Sr., R. O., 229
Machen, T. O., 28
Machen's Café, 31
Mack Memorial Library, 162
Mack, Sr., Dr. Donald G., 68, 229
Mack, Willie, 163, 229
Magnolia Petroleum Company, 53
Main Street Program, 163, 238, 244
Mall Pharmacy, 101, 102
Mann, J. L., 97
Mansfield, Louisiana, 41, 42
Manuel, Liberty "Frenchie," 216
Mardi Gras parade, 123
Martin, A. B. "Shooter," 215
Martin, C. C., 197
Martin, Clinton, 201
Martin, K. O., 97
Martin, Lonnie, 201
Martin, Mrs. Ever, 201
Martin, Rev. William, 141
Martin, Sam, 111
Mason, Rantz, 47
Mason, Wanda, 111
Masonic Lodge, 113, 118, 119, 227
Mathis, Peggy, x
Matthews, Mrs. Mary Will, 201
Matthews school, 200
McClain, Dr., 98
McConnell, Charles, x, 52, 53, 81, 213, 230
McCrae, John, 182
McDonald, Dr. Andrew Jackson, 15, 95, 230
McDonald, Rev. Peter, 201
McDonald, Sim, x, 241
McEachern, Darnell, 201
McEachern school, 200
McGill, S. S., 24, 140
McGowan, Frances, 133
McLain, Bobby, 112
McLaren, Rev. H. S., 137
McMahen, Barbara, 133
McMahen, Dr. Royce, 103
McMahen, Dr. Wayne, 99, 103
McMullan, Denny, 78, 163
McWhorter, Dr. John, 186, 187
McWilliams, Charles Terrell, 96
Melancon, Derek, xi
Methodists, 135, 136, 138, 201, 217
Miers, Butler, 123
Mill town life, 22
Miller, Arthur Logan, 68
Miller, Aubrey, 137, 138
Miller, W. A., 66, 235
Miller, William Alexander, 68
Mills, Olin, 101, 102
Milner, Frank, 215
Minden Bank and Trust Company, 52, 158
Minden Press-Herald, 196, 202
Mitchell, Dorice, 137
Mitchell, George, 215
Monroe, W. W., 217
Montgomery, Jim, 68
Montgomery, John M., 68
Montgomery, John M. "Jack," 122, 221, 231
Montgomery, June, 215
Montgomery, Pearl, 215
Monzingo, Samuel, 15
Moody, Bob, 142
Moody, Gay, 142
Moody, Howard, 102
Moore, Fannie, x, 116
Moore, Mae Dee, 69
Moore, Simon D., 197
Morgan, Ray, 48
Morgan, Ronald, 138

Morris, Dr. Marlin, 98
Morris, Ed, 196
Morris, W. M., 113
Mosely, Bill, 122
Mothershed, Mrs., 66
Mount Moriah Church, 197
Mouser, Adene, 115, 123
Mouser, Aubrey, 115
Mt. Paran Baptist Church, 201, 206
Murals, 163
Murph, Don, 111, 112
Murph, Jimmie N., 79
Murph, Jimmie Sue, x
Murphy, Reverend Randall, x, 143
Murray Memorial Baptist Church, 136, 224
Murrell, John, 18, 19
Murrell, Joseph, 15

Nail, L. D., 80
National Association for the Advancement of Colored People, 188
Nations Brothers Packing Company, 50, 58
Neal, Thomas, 15
Nelson, Linda, 112
Nesbitt, Mrs. Mattie, 142
Nevin, John, 25, 40, 42, 43
New Bethel A.M.C. Church, 148
New Bethel African American Methodist Episcopal, 138
New Bethel African American Methodist Episcopal Church, 217
New Bethel AME, 189
New Sarepta Baptist Church, 206
New Shongaloo School, 200
Newbourne, S. J., 182
Newsom Dairies, 200
Nichols, Roy, 122
Noles, Richard, x, 70
North Arkansas Street Church of Christ, 138
North Louisiana Moral and Civic Foundation, 143
North-South Construction Company, 192
North Webster Ambulance Service, 109
North Webster Industrial District, 164, 213, 237
North Webster Industrial Park, 167, 174
North Webster Parish Industrial District, 22, 25, 51, 228, 243
Northwest Louisiana Technical College, 77, 162
Norton Funeral Home, 63
Norton, Lester, 140
Nunn, Mrs. Era, 201
NZ Cash Store, 53

O'Bier Insurance, 238
O'Bier, Tommie, 17
O'Glee, James, 102
O'Neal, Rubin, 27, 80
Oakley, W. R., 224
Oden, W. H., 224
Oden, Waymon, 52, 53, 230
Odom, Dr. Bill, 112
Odom, Dr. William, 102
Office Dimensions, 172
Offutt, Paul, 53
Oil and Gas, 53
Old Sarepta Missionary Baptists, 204
Old Shongaloo Missionary Baptist Church, 201
Old Shongaloo School, 200
Old Shongaloo United Methodist Church, 201
Old Union Baptist Church, 201
Olive, Ed, 122
Orange, Catherine, 112

Pace, Mac, 122, 220
Pacey, Dr., 101
Padget, Louis, 69
Pafford, Leon, 195
Palace Radio Shop, 53
Panic of 1873, 177
Park, Charles, 112
Park, Wm. Charles, xi
Parker, Cordell, 216
Parlor games, 123
Payne, Dr. Charles, 98, 112, 118
Payne, Marietta, 112
Payne's Pressing Shop, 162
Peaden, Rev. Carol, 195
Pearce, J. M., 97
People's Bank and Trust Company, 52
Percy Cobb Dam, 127
Perkins, Arthur, 24
Perkins, Rev. Jesse, 142
Perritt, Careece, 225
Peters, Bobbie, 111
Pharr, Louise, 111
Piccadilly Supper Club, 215, 218
Pick and Pay Grocery, 53
Pickard, Ernest, 115
Pickett and Sons, 53
Pickett Enterprises, 55
Pickett, G. B., 215
Pickett's Store, 55
Pilgrim's Rest Church, 201
Pine Plaza, 171
Pine Woods Lumber Company, xiii, 21, 23, 24, 27, 51, 66, 95, 138, 141, 157, 162, 163, 178, 212, 230, 234
Pioneer school, 200
Pipes, Abraham, 8, 15
Pixie Family Store, 200
Pixley, Ray, 80
Plauche, Rev. M. L., 139
Plunkett, Gene, 180
Police Jury, 83, 96, 222
Pompey Flowers, 213

Pope, Mr., 66
Porter, Solomon, 80
Posey, Rev. Ralph, 139
Potters school, 200
Powell, Alvin, 137
Powell, Larry, 137
Presbyterians, 135
Princess Café, 53, 215
Provost, C. J., xi

Rainey family, 8
Rancho Drive-In Theater, 59
Reeder, Benny, 123
Reeves, Rev. Wayne, 142
Regions Bank Corporation, 52, 159, 196, 198
Relay Station, 123
Renner, Bob, 112
Revelle Studio, xi
Reynold's Store, 53
Reynolds, Bertha Smith, 235
Reynolds, G. I., 26
Rhone, Cherokee, 68
Rhone, Henry, 189, 216, 231
Rhones Dry Goods, 196
Rhynes, Betty, x
Rhynes, Bettye Benton, xi
Rice Copeland's Store, 200
Ridgel, Malachi, 189
Riggs, Johnnie, 133
Riis, Erling, 24, 38, 39, 121, 232
Riley, Cora, 29
Ritz Café, 32
Robert Max Hayes barbershop, 200
Roberts, Patsy, 111
Robertson, Dennis, 99
Robertson, Dr. Jason, 101
Robertson, Dr. Raymond, 99, 102
Robertson Family Pharmacy, 101, 102
Robertson, James, 99
Robertson, Jimmy, 99
Robinson, Dr. Archie, 98
Robinson Drug Store, 53, 101
Robinson, J. A., 198
Robinson, Steve, 80
Rock Hill, 66
Rocky Hill, 67
Rodeo, 123
Rodeo Arena, 128
Rodeo parade, 123
Roemer, Senator, 213
Rogers, Jack, 68
Roman Catholic Church, 135
Roney, C. P., 136
Roseberry, William B., 200
Roseberry's store, 200
Roseburough, John, 69
Rotary Club, 114, 220
Rowland, J. B., 15
Rowland, Jim, 180
Rowland, Mrs. A. B., 17
Roy's Pharmacy, 101
Rukeyser, Louis, 159
Russell's Pharmacy, 162
Rutledge, Dr. J. E., 101

Sacred Heart Roman Catholic Church, 139, 153
Salim, Emmitte, 52
Sams Mens Wear, 196
Sanders, A. M., 137
Sanders, J. A. "Junior," 122
Sanders, Marshall, 140
Sanitary Dairy, 50, 61, 192
Sarepta High School, 197, 205
Sarepta, Louisiana, 194, 196, 197
Sawyer, Olene, 116
Sears, "Babe," 129
Segregation, 188
Senior Center, 116
Senior Friends, 114, 119, 220
Sessions, Dr. Wayne, 98, 217
Sessions, Jerry Wayne, 68, 233
Sexton, C. C., 201
Shady Grove Methodist Church, 201
Shakalo, 199
Shirley, Joyce Friday, 116
Shongaloo Grocery, 200
Shongaloo High School, 200, 221, 239
Shongaloo, Louisiana, 194, 198, 199, 200, 202, 208, 230, 236
Shongaloo School, 205
Shongaloo United Methodist Church, 203
Shope, H.D., 24
Short, Mrs., 215
Shultz, Ed, 26, 81, 86, 52, 96, 129, 211, 214, 232
Sikes Ferry, 201
Sikes Ferry school, 200
Sikes, J. S., 198
Sikes, Jesse, 200
Sikes, John Turner, 199
Simmons, Eric, 115, 123
Simmons, Evelyn, 162
Simmons, Greg, 99
Simmons Rodeo Co., 115
Sims Clinic, 97
Sims, Dr. Howard, 97
Sims family, 8
Slack, Allie Mae, 201
Slack, B. L., 52
Slack, Barry, 52
Slack, Bobby, 52
Slack, C. A., 97
Slack, Ira Benton "I. B.," 52, 216, 233
Slack, J. B., 52
Slack, O. M. "Boss," 42, 234

Slack, Paul, 115
Slack, Robert, 201
Slack school, 200
Slack, Waymon, 115
Slack, Wilburn, 223
Slattery, Jr., John, 52, 79
Slavery, 187
Smith, Dorothy, 41
Smith, Dr. Wayne, 103
Smith, George, 6
Smith, Gertrude, 69
Smith, Henry, 115
Smith, Howard, 121
Smith, N. B., 113
Smith, R. A., 26, 38
Smith, R. A. "Buck," 138, 234
Smith, Raydell, 68
Smith, Robert Charles, 68, 138, 181, 185, 235
Smith, Sherrel, x, 79, 80
Smith, Suzanne Souter, 102
Smith, William, 114
Smith, William Byrd, 68, 235
Smith's South Central, 166
Soil Products Incorporated, 168
Soileau, Dr. Marvin, 98, 99, 100
Soileau, Jean, 111
Souter, Ted, 68, 215
South Main Mall, 173
Southern Methodist Church, 140, 145
SPARC Center, 207
Spring Theater, 63, 234
Springhill Airport, 91
Springhill Art League, 114, 119
Springhill Bank and Trust Company, 52, 60, 64, 159, 227, 232, 234, 237
Springhill Christian Church, 139, 149
Springhill Church of Christ, 136
Springhill Community Activities Center, 189
Springhill Community Center, 237
Springhill Country Club, 126
Springhill/Cullen Chamber of Commerce, 216
Springhill depot, 18, 20
Springhill dot Net, xi
Springhill General Hospital, 98, 99, 107
Springhill High School, 67, 70, 71, 74, 114, 120, 122, 221, 229, 236
Springhill Ice Plant, 89
Springhill Junior High School, 67, 239
Springhill, Louisiana, xiii, 17, 18, 25, 43, 78, 163, 188, 194, 213, 220, 221, 229, 236, 243, 244, 245
Springhill Lumber Company, 50
Springhill Medical Center, 99, 108, 233
Springhill Medical Services, 99
Springhill Missionary Baptist Church, 139, 154
Springhill Motor Company, 64, 170
Springhill Presbyterian Church, 140, 145, 230
Springhill Press and News Journal, 101, 123
Springhill Quarterback Club, 114
Springhill Riding Club and Rodeo, 115
Springhill Sports, 121
Springhill Telephone Company, 53, 57
Springhill United Methodist Church, 141, 146, 223, 227, 234
Stampley, Joe, 68, 235
Standard Oil Wholesale, 198
Stanford, B. B., 137
Stanford, Sammy, 111
State Theater, 57, 237
Stauffer Chemical, 50
Steel, Robert, 198
Steiner, Tommy, 115
Stephens, Diane, x
Stephens, J. D., 113
Stephens, Jerry, 80
Stevens, Mrs. Betty, 142
Stewart, Lloyd, 105
Stewart. W. E., 97
Stiles, Zolen, 68
Strater, Mrs. Verna, 162
Strickland, J. L., 113
Strother, Roy, 112
Sunbridge Healthcare Corporation, 100
Swan Clinic, 105
Swan Memorial Hospital, 97
Sweet Home Baptist Church, 201
Sykes, John Turner, 14

Tanner, Dr. Charles, 101
Tatum, Kay, 114
Taylor Academy, 67
Taylor, George, 115
Teague, Dr. Don, 99
Telephone Exchange, 198
Temple Baptist Church, 142, 151
Tenneyson, Murray, 52
Tennyson Drug Store, 26, 53, 62, 101, 215
Tennyson, Murray, 101
The Hill, 224
"The Old Hickory Tree", 197
The Old Oak Tree, 198
The Ritz, 216
Thigpen, Ed, 121
Thomas, Dale, 118
Thomas, Daniel G., 83
Thomas, Jimmy, 83, 189, 236
Thompson, Jerry Dale, 99
Thompson, Leo, 115
Threet, Richard, 99
Timothy School, 67
Torrence, Dr. Gary, 98, 99
Trane Company, 167
Treat, F. B., 97
Trees, vi
Trinity Chapel, 142

Trinity Worship Center, 142, 149
Troquille, Billy, 215
Tucker, Marvin, 215
Tulane University, 236
Tullett, John, 48
Turgeon, Romeo, 215
Turner, Dexter, 196
Turner, Dr. Woodrow, 68
Turner, James, 68
Turner, Junior, 122
Turner, Woodrow, 164
Twitty, Billy, 115
Tyler, Charles E., 68
Tyler, D. G., 52
Tyler, George, 198
Tyler, Roy, 80

Uncle George's Café, 212, 217
Union Springs Baptist Church, 201
Union Springs School, 200
United Pentecostal Church, 154
University of Georgia Press, xi
University of North Carolina at Chapel Hill, 236
University of the South in Sewanee, Tennessee, 236

Vargin, Jennifer, 196
Vaughn, J. C., 137
Vaughn, Rev. R. W., 141
Vines Insurance Agency, 200
Visitor Information Center, 238

Wal-Mart, 168
Wall, Dr., 98
Wallace, John, 123
Walnut Road Baptist Church, 151
Walnut Road Missionary Baptist Church, 142
Ward, George, 24
Washington, Bobby, 196
Washington Church, 189
Washington Church of God in Christ, 143, 149
Washington, Elder A. D., 143
Washington, Jonathan, x, 112, 138, 189, 237
Waters, Eugene, 26, 52, 54, 237
Webb, Rev. Floyd, 195
Webster Council on Aging, 116
Webster Parish, 197, 199
Webster Parish Library, 92
Webster Parish Police Jury, 41
Webster Parish School Board, 41, 188, 189, 201, 224, 227, 230, 235
Webster Parish Schools, 229
Webster Parish Tourism Commission, 164
Webster Theater, 59, 211
Western Auto Associate Store, 53
Whipple, Jr., James W., 182
White, Floydean, 196
White, James Milton, 180
Williams, Alicia, 133
Williams, Dewey, x, 54
Williams, Elder Edward, 143
Williams, Elder Errol, 195
Williamson, Dr. Samuel Ruthven, v, xi, xiv, 68, 236
Williamson, Page, x, 41, 42
Williamson, Sr., Mrs. S. R., 246
Williamson Sr., Sam, 47
Willie Mack Department Store, 53
Willis, Darrell, 140
Willis, Harvey, 112
Willis, Jan, x, 82, 163
Willis, Janis Stroud, 237
Wilson, Rev. Randy, 137
Wilson, Wilburn "Will," 113, 238
Wimberly, D. C., xi, 181, 185, 239
Wise, Dr. Giles James, 199, 200
Wise, Dwayne, 215
Wise, Elton, 201
Wise, James, 15, 199
Wise, Rodney, 201
Wise, Roger, 201
Wise, Roy, 182
Wise, Ruby, 111
Woodchoppers, 116
Woodmen of the World Building, 113
Woodson, Rev., 195
Wooley, Steve, 198
World War I, 179
World War II, 180, 239, 244

Young, Dr., 52, 101
Young, Mertis, 200
Youngblood, Rev. John, 141